D0009191

THE GUIDE TO EXTRAORDINARY CURIOSITIES OF OUR UNIVERSE

WONDERS OF THE NIGHT SKY
YOU MUST SEE BEFORE YOU DIE

BOB KING

CREATOR OF ASTRO BOB AND AUTHOR OF *NIGHT SKY WITH THE NAKED EYE*

PAGE STREET
PUBLISHING CO.

PAGE STREET
PUBLISHING CO.

First published in 2018 by
Page Street Publishing Co.
27 Congress Street, Suite 105
Salem, MA 01970
www.pagestreetpublishing.com

Distributed by Macmillan, sales in Canada by The Canadian Manda Group.

22 21 20 19 18   1 2 3 4 5

ISBN-13: 978-1-62414-492-9
ISBN-10: 1-62414-492-6

Library of Congress Control Number: 2017959350

Cover and book design by Page Street Publishing Co.

Printed and bound in the United States

To my mom and dad,
Lorraine and Bill, for their love
and encouragement.

CONTENTS

Introduction

Life. I'm a firm believer in living it fully. When it comes to my favorite hobby, astronomy, I strive to witness as much of the cosmos as lack of sleep will allow.

Each day, the inscrutable blue dome of sky greets us when we step out the door. What's up there? Only the entire universe beyond our little blue spaceship called Earth.

Beauty runs deep overhead. Deep as in billions of light-years both in space and in time. Every moonrise, every heart-stopping meteor and every wish upon a star plays out up there. There's so much to see.

But what to pick and how to see it? Allow me to lend a helping hand. Lots of us have bucket lists or things we want to do before we log off from this world for good. Float in the Dead Sea, ride a double-decker bus in London, eat truffles or learn to ballroom dance. We want to live life to the max, filling our buckets to the brim.

You may not be able to get to the Grand Canyon or the Great Wall of China, but when it comes to the sky, you came to the right place.

Have you ever wanted to see a total eclipse of the Sun? A magnificent, sky-brimming aurora? Or how about Saturn's rings with your very own eyes?

In this book, we'll look at 57 must-see sights in the night sky you won't want to miss. Some will require travel, others the use of a telescope or binoculars, but many can be experienced with little effort and zero expense.

In choosing the entries, I carefully evaluated what I thought made for a magnificent viewing experience or provided an opportunity to get to know an essential aspect of the cosmos—like seeing an asteroid or the closest star system beyond the Sun, Alpha Centauri.

Any list is selective, so I hope I didn't pass over one of your favorites. Some, like the gegenschein and Winter Hexagon, were left out because I covered them in my first book, *Night Sky with the Naked Eye*. Others, I've revisited in this book in more detail.

Bigger than anything else in human experience, the sky and the wonders it holds invites us to deepen our sense of appreciation for the world and encourage a truly cosmic perspective on life. Let the bucket filling begin.

To Russ,
Thanks for having me. I hope you get to see some of these wonderful sights!
Bob King

Magnificent Saturn

Don't let life slip by without seeing one of nature's most remarkable sights, the quintessential emblem of all things outer space. Ignore all the rest if you must, but see Saturn.

You'll be hard-pressed to find anything else like it: a ball surrounded by levitating rings. Thousands of them nested one within the other and composed of dusty water–ice, ranging in size from sand grains to boulders. The planet averages about 800 million miles (1.3 billion km) from Earth, a distance so great, their gritty texture blends into a creamy smoothness.

↑ *Saturn is the quintessential must-see night sky sight. A small telescope and knowing where to look are all you need. Credit: Damian Peach*

7

In a pair of binoculars with a magnification of 10x, Saturn looks oval-shaped because the rings blend with the ball of the planet. But any small telescope magnifying 30x or higher will clearly show the globe of Saturn nestled within its rings, like a baby in a mother's arms. The most common reaction to seeing Saturn for the first time is that it doesn't look real. Photos of the other celestial bodies taken with large telescopes rarely resemble what you can see with your eye, but that's not the case with Saturn. The view, while on a smaller scale, uncannily resembles what the camera sees. No wonder we can't quite believe it's real. Yet there it is. Real as rain.

Higher magnification may reveal the most prominent gap in the rings called **Cassini's Division**, which has less ice than some of the other rings and so appears dark in contrast. Cassini's separates the wide B ring from the narrow A ring.

Saturn orbits the Sun once every 29.5 years, and its axis is tilted 27°, much like Earth's. This causes the rings to change orientation over time. Sometimes we see the north face, other times the south, and for a brief time every thirteen to fourteen years, the rings are edgewise and almost disappear from view.

Disappear? Yes! The main rings extend 180,000 miles (290,000 km) end to end, so the planet would comfortably fit between the Earth and the Moon with 60,000 miles (96,000 km) to spare. But, as wide as they are, they're incredibly thin—only about 30 feet (10 m) thick if you don't count warps and ripples.

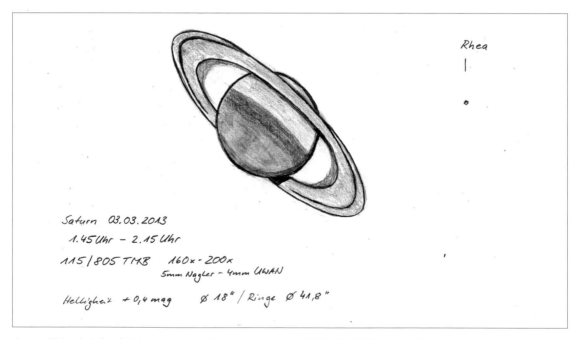

▲ *This sketch of Saturn was made using a small 4.5-inch (115-mm) refractor and magnification of 160x-200x. It shows how much detail is visible in good seeing conditions, including the bright A and B rings and fainter inner C ring, Cassini's Division and a broad equatorial cloud belt. Credit: Uwe Schultheiss*

How thin is that? A sheet of 8½ x 11 inch (21.6 x 27.9 cm) U.S. writing paper has a thickness-to-length ratio of 0.00036. To scale, Saturn's rings are 10,000 times thinner! Much thinner than the thinnest razor blade.

All of that dirty ice may once have belonged to small moons that collided in the relatively recent past; their fragments spreading out to form the rings. Or an icy comet may have struck an orbiting moon and blasted it to pieces. The origin of the rings remains a mystery, but their beauty can be enjoyed by anyone with a small telescope. Dig yours out of the attic, then use one of the free apps or home planetarium programs listed on page 10 to find out when and where Saturn will shine in your night sky.

Careful, though. One look at the planet could change your life forever, so do not go gently to that eyepiece. You might get bitten by the astronomy bug and soon find it hard to get to bed at a decent hour on clear nights. I speak from experience.

How to find Saturn

The diagram shows Saturn's location one month past **opposition**—closest approach to Earth—when it's conveniently placed for viewing in the evening sky. Saturn is bright and stands out from all of the stars in the half-dozen constellations it will pass through between now and the late 2020s.

Even easier, download a star chart app for your phone, point it skyward and let the app do the finding. When to look? Through the mid-2020s, Saturn will be best visible during evening hours from July through November.

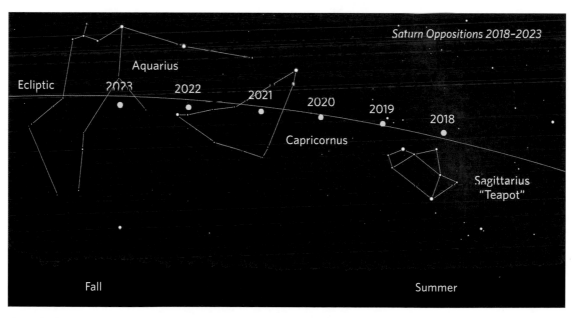

▲ *The best time to look for Saturn is around the time of opposition, when it's closest to Earth and brightest. This diagram shows the planet's location at five different oppositions as viewed from mid-northern latitudes (40° North) in the southern sky during evening viewing hours. Source: Stellarium*

How about a telescope?

A telescope is a must for appreciating the unique beauty of Saturn and several of our other must-sees. If there's an astronomy club in your area, chances are that the members would be more than happy to let you know when they're holding their next public observing night.

If you decide that Saturn alone is reason enough to finally invest in a telescope, I recommend an affordable 4- to 6-inch (100- to 150-mm) reflecting telescope on a Dobsonian mount. The name comes from John Dobson, who devised a mount that was simple, stable and sturdy. You push and pull the scope up and down then right and left to center your target. Easy, right? Many inexpensive telescopes come with wobbly mounts and tripods. Not a Dob.

That said, even junky scopes will show the rings, but why get junk? Invest a little cash and you'll be able to tour the universe in comfort and style! In the resources section (page 216), I've listed several telescope options and online shops where you can buy with confidence.

RESOURCES

- To find where Saturn is in the sky, Google "Star Chart app." There are many fine programs for both iPhone and Android that will pick up on your location and show the sky any time of day or night. Several are free, including Star Chart:

- FOR IPHONE:
  itunes.apple.com/us/app/star-chart/id345542655?mt=8

- FOR ANDROID:
  play.google.com/store/apps/details?id=com.escapistgames.starchart&hl=en

- You can also try the versatile Stellarium app ($2.99).

- FOR IPHONE:
  itunes.apple.com/us/app/stellarium-mobile-sky-map/id643165438?mt=8

- FOR ANDROID:
  play.google.com/store/apps/details?id=com.noctuasoftware.stellarium&hl=en

- Or download Stellarium for free on your laptop (Mac or PC) from stellarium.org.

Moon–Planet / Planet–Planet Conjunction

Some things in life are inevitable. It *will* snow in Duluth, Minnesota, in January. People *will* disagree about politics. And given time, two planets *will* come together in conjunction to make a pretty diadem in the night sky.

A **conjunction** occurs when one or more celestial objects lines up above another. We say they have the same celestial longitude. Remember longitude? Those are the vertical lines on a world map. A city with the same longitude as yours is located either due north or due south of your location. For instance, St. Louis's longitude is nearly identical with Jackson, Mississippi, though the two cities are almost 500 miles (804 km) apart. You could say they're in permanent conjunction. In contrast, celestial alignments last only a day or two.

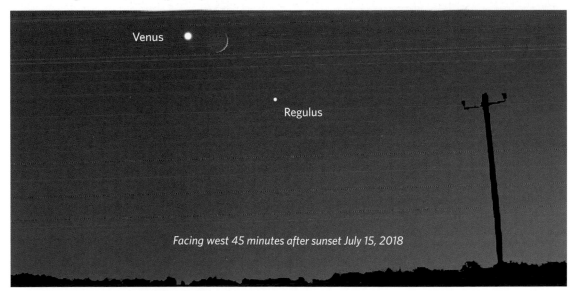

Facing west 45 minutes after sunset July 15, 2018

▲ *Venus, the waxing crescent and Regulus gather for an eye-catching trio on July 15, 2018. Source: Stellarium*

Planets never get stuck in the same place in the sky because they're always on the move, orbiting the Sun. They also travel at different speeds depending on their distance from it. Mercury, closest to the Sun, completes one lap every 88 days, Venus every 225 days and Jupiter every twelve years.

Further, each planet's orbit is tipped somewhat with respect to Earth's. Mercury's inclination is the greatest of the eight planets at 7° (a stack of about fourteen full moons), but most are closer to 2 to 3°. In the big picture, these variations are slight: All the planets travel within a narrow band through the twelve constellations of the zodiac. We'll always find them there and nowhere else. Never in Orion, never in the Big Dipper, never in the Southern Cross.

Their restricted movement stems from the very origin of the solar system. The Sun and planets originated from a large cloud of stardust and hydrogen gas dubbed the **solar nebula**. Gravitational collapse of the cloudlet 4.6 billion years ago led to a hot, concentrated core that became the Sun. The remainder of the mass spun out into a disk that coalesced into the familiar planets, asteroids and comets.

If you start with a gas–dust cloud with just a little bit of spin, something fascinating happens as it becomes compressed by self-gravity: It spins faster and faster! Physicists call it the **conservation of angular momentum**. You can see the principle at work during an ice-skating competition. When a skater goes into a spin, she pulls in her arms and a leg—becoming more "compact" as it were—and starts spinning so fast it makes you dizzy just to watch. When this very same process happened in the solar nebula, the energy of spinning flattened out the cloud into a pancake-like disk where the planets formed.

▲   *Venus and the crescent Moon meet in a beautiful conjunction at dusk in the western sky. Whenever Venus is visible at dusk or dawn, the crescent Moon is a regular visitor, passing near the planet about once a month. Credit: Bob King*

That relic remains with us to this day. Because Earth also resides in the same plane as the other planets, when we gaze into the sky, the planets cycle around the sky in the same narrow belt. Coming and going, they appear to pass near one another in close and awe-inspiring conjunctions.

Moon–planet conjunctions are fairly common because the Moon is close to Earth and moves across the sky much faster than the distant planets. Over the course of a month, the Moon "drives by" all eight planets, spending a night with each. Some of the most striking conjunctions occur when the lunar crescent passes near the brightest planets, Venus and Jupiter. The sight of two brilliant celestial objects side by side can get the attention of even those who don't routinely watch the sky.

Venus

Mercury

Jupiter

Facing southeast 1 hour before sunrise Dec. 21, 2018

▲ *Jupiter and Mercury pair up with Venus a little more than two fists away on December 21, 2018. Source: Stellarium*

Planet-planet conjunctions are uncommon and happen over several nights because the planets are so much farther away and appear to move much more slowly across the sky. They gradually grow closer until the night of conjunction, when they're closest and most pleasing to the eye.

No planet stands still. As they come together, so they part. For the more distant planets, which take longer to orbit around the Sun, they don't meet up again for years. Mercury and Venus, being much closer to Earth, zoom back and forth from morning to evening, often crossing paths with each other or with the more remote planets.

Not surprisingly, the most visually compelling conjunctions involve the brightest planets, Venus and Jupiter, or one of these paired with Mars, when that planet is close to Earth and brilliant.

Conjunctions are the sky's magic tricks. Two planets as distant from one another as Venus and Neptune can briefly look like close neighbors when they happen to fall within the same line of sight even though they're more than 2.5 billion miles (4 billion km) apart.

Not only are planet pairings wonderful to look at, we can use them to help us identify planets invisible to the naked eye, such as Uranus and Neptune, both of which require binoculars.

While Uranus can be glimpsed with the naked eye from a dark sky, having Jupiter or Mars lined up nearby in conjunction makes finding it no trouble at all. Otherwise, you'll need a fairly detailed map and knowledge of the zodiac constellations to spot them.

Facing southeast 45 minutes before sunrise Jan, 22, 2019

▲ *Three planets and Scorpius's brightest star, Antares, cluster together in the east before sunrise on January 22, 2019. Source: Stellarium*

Very rarely, planets pass so close they overlap or **occult** one another. I've listed several below—they're few and far between, so pass the news to your kids and tell them to pass it on to their kids.

You may hear or read online that planetary alignments are responsible for everything from plagues to earthquakes. Don't believe it for a minute. While there is additional gravitational attraction on Earth when two or more planets line up, it's negligible because they're so far away. Venus briefly comes closest to the Earth at 27 million miles (43.5 million km) when it swings between the Earth and the Sun. That's 113 times farther than the Moon. The tides caused by the combined tugs of the Sun and Moon are much stronger than Venus's minute contribution, and there is no correlation during those approaches with earthquakes and other natural disasters.

How to see two planets line up

There are lots of ways to find out when two planets are in conjunction. The *Old Farmer's Almanac*, published annually since 1792, lists all planetary conjunctions in its monthly calendar section. You can also download a free star program such as Stellarium to thumb through the sky days, weeks and months ahead and see for yourself when planets approach and recede from each other.

Or you can use the table on the next page, prepared through 2223, listing the best and brightest conjunctions. If you're looking for bright-faint planet pairings, please turn to the Uranus and Neptune entry on page 139.

Moon-planet conjunctions

This sample for 2018 will get you started. For a more complete listing, check the resources section at the end of this chapter (page 16).

Separations between the Moon and planet are given in degrees. One degree is equal to two full moons or the tip of your little finger held at arms length against the sky. Dates and separations are what North American observers will see.

- June 27, evening: Saturn 0.75° south of the Moon
- July 14, dusk: Mercury 2° south
- July 15, dusk: Venus 1° south
- July 24, evening: Saturn 1.5° south
- September 13, dusk: Jupiter 3° south
- October 11, dusk: Jupiter 2.5° south
- October 14, evening: Saturn 1.5° south
- November 15, evening: Mars 2.5° north of the Moon
- December 3, dawn: Venus 5° south
- December 14, evening: Mars 3.5° north

Planet-planet conjunctions

2018 TO 2019

- December 21, 2018, dawn, low in the southeastern sky: Mercury 1° north of Jupiter
- January 23, 2019, dawn, low in the southeast: Venus 2.5° north of Jupiter
- February 18, 2019, dawn, low in the southeast: Venus 1° north of Saturn
- November 24, 2019, dusk, low in the west: Jupiter 1.5° north of Venus
- December 11, 2019, dusk, low in the west: Saturn 1.75° north of Venus

Planet occultation table

These are rare!

- Venus occults Jupiter on November 22, 2065, but they'll be too close to the Sun to observe for most telescopes.
- Venus occults Jupiter again on September 14, 2123, but it occurs over the Pacific Ocean. Get your cruise booked early!
- Mercury occults Mars on August 11, 2079, seen from the Middle East at sunrise.
- Mars occults Jupiter on December 2, 2223, in the early morning hours across the Americas.

RESOURCES

- *Old Farmer's Almanac*: (available from Amazon and Barnes & Noble) and published in late summer prior to the upcoming year
- List of upcoming conjunctions: en.wikipedia.org/wiki/List_of_conjunctions_(astronomy)#2018
- Stellarium or a phone app like Sky Chart described for Saturn on page 10. With these programs, you can advance the time to scan the future (or past) and find out what's in store.

3

Three-Planet Conjunction

While two-planet pairings aren't uncommon, seeing three glimmering together in a small corner of sky is as rare as winning big in Vegas. But if you're prepared, I guarantee this jackpot's attainable. Because the movements of the planets are known with great precision, it's simply a matter of keeping track of upcoming conjunctions and making sure you get outside during the several nights the planets will be close together.

I should be honest at the outset and say that the chances of three planets lining up exactly one atop the other is, well, astronomical. That's why we'll call a spade a spade and identify these triple events as exactly what they are—**dual-pair conjunctions**. These occur when three planets gather so closely together that two close conjunctions occur within days of each other. For instance, in 1991, Mars was in conjunction with Jupiter followed four days later by a Jupiter–Venus conjunction with Mars still close by. The planets practically tripped over each other's feet.

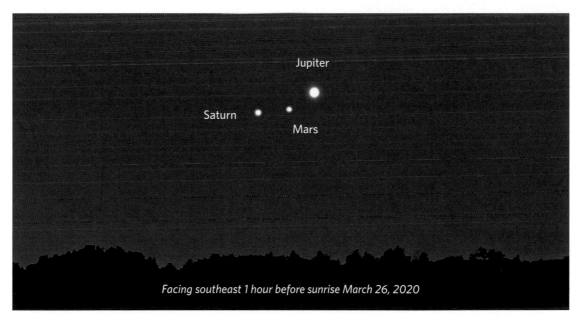

Facing southeast 1 hour before sunrise March 26, 2020

▲ *Saturn, Mars and Jupiter form a compact group at dawn on March 26, 2020. Source: Stellarium*

When three planets gather, typically at dusk or dawn with Venus or Mercury involved, their individual motions create an ever-changing series of triangular groupings that are delightful to watch as they shape-shift from night to night.

One of the last really nice triple-planet conjunctions occurred in mid-June 1991. On June 17 that year, Venus, Mars and Jupiter crowded into just 1.8° of sky, an amount easily covered by an extended thumb. Two nights prior, the crescent moon joined the scene for a memorable evening twilight sight. Little did I know at the time that the previous triple crush had happened 90 years earlier when Mars, Jupiter and Saturn banded together on November 28, 1901. Holy cow, even my grandparents were too young to see that one!

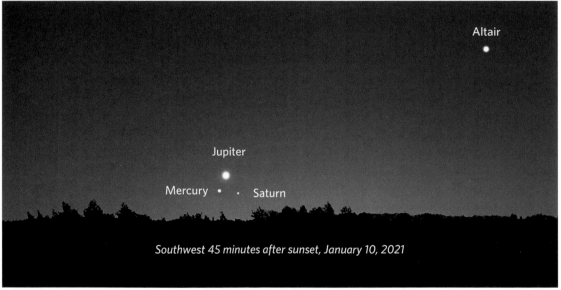

Southwest 45 minutes after sunset, January 10, 2021

⋏ *Jupiter, Mercury and Saturn gather into a triangle only 2° on a side on January 10, 2021. Source: Stellarium*

⋏ *Venus (right), Jupiter (bottom) and Mars crowd into a triangle less than 2° on a side to make a dazzling sight on June 17, 1991. Copyright: Alan Dyer/AmazingSky.com*

Tight three-planet groupings are fairly rare, but several are visible through the early 2020s. The more compact the group, the more infrequent the occurrence, so to give us a shot at seeing one, I've expanded the definition to include planets that pass within several degrees of each other. While not necessarily in conjunction, they'll be close enough to one another to make an eye-catching sight.

Though a four-or-more planet conjunction may be possible, the exceeding rarity means waiting many generations. We might find satisfaction with a substitute event—not a conjunction exactly, but more of a line-up of planets across one section of sky. While uncommon, these occur more frequently than compact triples and are well worth seeking out. To find out more about how and when to view one, see page 95.

How to see a triple-planet conjunction/planet gathering

Because they can last several days, the odds of beating the clouds increase, so you can usually catch one of these triple treats on at least one night. I've peered into my crystal ball software to find opportunities through 2022 and listed them below. Be patient and hope for clear skies—you'll see one soon enough.

Close three-planet groups

- March 25, 2020, start of dawn, low in the southeastern sky. Linear grouping of Saturn, Mars and Jupiter in a line 6.5° long, or equal to a little more than three fingers held at arm's length.
- January 10, 2021, dusk, very low in the southwestern sky about 40 minutes after sunset. Jupiter, Mercury and Saturn in a tight triangle only about 2° on a side. Use binoculars.
- February 25 to March 5, 2021, dawn, very low in the eastern sky about 40 minutes before sunrise. Mercury, Jupiter and Saturn within about 7° of each other. Use binoculars.
- March 24 to April 5, 2022, dawn, low in the southeastern sky. Venus, Saturn and Mars will form a triangle about 5° on its longest side that changes shape with each passing morning as the planets move along, each at its own rate of speed. A thin crescent moon joins the gang on March 28.

RESOURCES

- Bright planetary conjunctions and other unusual alignments: climate.gi.alaska.edu/Curtis/astro5.html
- Mutual planetary occultations: bogan.ca/astro/occultations/occltlst.htm
- Stellarium or a phone app like Sky Chart described on page 10.

Summer Milky Way

There are two Milky Ways: winter and summer. The summer version is by far the brighter and easier to see. It's also the time of year we're more likely to be outside at night. January's Milky Way looks anemic in comparison but crosses the sky in identical fashion as its July counterpart. I love the sight of both and look forward to their return every year.

When someone asks where you live, you give them your address and perhaps the city and state, too. But a more complete answer might go something like this: country, continent, planet, solar system, Orion Arm and Milky Way Galaxy. We could keep on going—Local Group, Virgo Cluster, Laniakea Supercluster, universe—but let's stop at the galaxy level, a place big enough to rock your mind with the enormity of it all.

Dream big, right? That's what they say. Well, no dreaming needed here. The Milky Way, seen from outside, looks like a pointillist pinwheel dotted with nearly 400 billion stars. Some gather in a central bar-shaped feature in the galaxy's hub, others shape the spiral arms and flattened disk while outliers define the limits of the galactic halo. The whole work spans 100,000 light-years, a distance so unimaginably vast that it takes 100,000 years for a beam of light to travel from one end of the galaxy to the other. Even if we shrunk the Earth to the size of a grape, the galaxy at the same scale would still be 55 billion miles (88 billion km) wide or nearly twelve times the distance to Pluto.

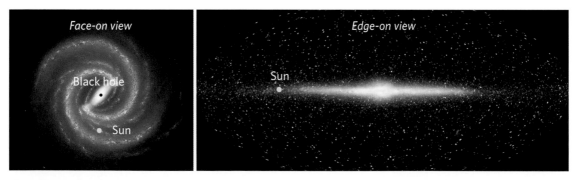

⌃ The Milky Way Galaxy is a multi-armed spiral with between 100 and 400 billion stars plus countless nebulae, star clusters and planets. Our Sun is located a little more than halfway between the galaxy's edge and center, home to a black hole with a mass more than four million times the Sun's. Source: NASA with additions by the author.

Our solar system resides in one of the spiral arms, a minor one called the **Orion Arm**, about halfway or 26,000 light-years from the galaxy's center, where a supermassive black hole, 4.5 million times more massive than the Sun, bides its time like a spider at the center of a web waiting for stray stars and asteroids to pass within its gravitational grasp. In summer, we look across the inner spiral arms and toward the star-rich central bulge, the reason the Milky Way band appears brighter then. In winter, we face outward across the outer arms to where the stars ultimately thin out and give way to virtually starless, intergalactic space. Without the density and number of stars to bulk up the view, the winter Milky Way band lacks the intensity of summer's. For northern hemisphere skywatchers, summer's profligate life and winter's dearth of it find their counterparts in the two Milky Ways.

If we live in a giant galaxy, not only made of stars but sprinkled liberally with interstellar dust, threaded by immense clouds of cold gas and populated by billions of planets and trillions of comets and asteroids, why doesn't it simply fill up the sky? Instead, we see a cloudy ribbon of light.

The solar system is embedded in the middle of the galaxy's flat disk, where, along with the bulge, most of the Milky Way Galaxy's stars are concentrated. When we gaze at the band of the Milky Way, we peer *into and through* the disk. Billions of stars across thousands of light-years lie along our line of sight. Most are too faint to see individually; instead, these pinpoint suns blend together to form a misty band of light that to first-time viewers resembles an arc of clouds. But make no mistake. Those "clouds" are comprised of billions of suns.

If you follow the band around the sky, you're looking through tens of thousands of light-years of stars, much like standing at the edge of a forest and looking in. You can make out the foreground trees but the more distant ones fill in every bit of open space until it's nothing but a dense thicket.

If we look up or down *out* of the galactic plane, our gaze passes through only about 5,000 light-years' worth of stars and then enters the emptiness of intergalactic space. The stars thin out in a hurry; there aren't enough of them to create a band. Instead, we see them randomly scattered across the sky. All the stars you see on a clear night belong to our galaxy just as a visitor to an alien planet in the Andromeda Galaxy would only see stars belonging to that galaxy.

Because we're stuck on Earth, we can only see half of the Milky Way band at any time. The other half is blocked by the horizon. But if you could jump into a spaceship and zoom away from Earth, say between here and the Moon, what a sight would await. Safely tethered to your spacecraft on a spacewalk, you could roll your body this way and that and follow the milky band with your eyes for all 360 of its degrees. Talk about amazing. In space, there are no Milky Way seasons.

But the view's not too bad down here on Earth. On a late July night, when the Milky Way flows like a fuzzy, phosphorescent river from horizon to horizon, all you need to do is look up and take it in. No thinking required. Few sights are more magnificent or impart such a grand sense of scale. Against this starry yardstick, we might measure the arrow of our lives, reflecting on how we arrived at this place and where we hope to go.

Sometimes, you need a whole galaxy to get a handle on this crazy thing called life.

How to see the summer Milky Way at its best

While the summer Milky Way can be viewed on late winter mornings before dawn or the winter Milky Way before dawn in late summer, we'll stick to convenient evening hours. For a full appreciation, plan your viewing session around the time of a new moon, so moonlight won't compromise starlight. The table below lists the dates of upcoming new moons; three or four nights either side of new moon is fine.

If you live in the countryside where light pollution isn't an issue, you're all set, but most of us will need to drive to a dark sky to soak in the galactic grandeur.

In the resource section on page 23, I've included a light pollution link. There, you'll find a map of towns, roads and highways with color-coded overlays that show where light pollution is least and greatest. Select a place within driving distance or, better yet, plan a late-summer excursion to a state or national park, located far from city lights.

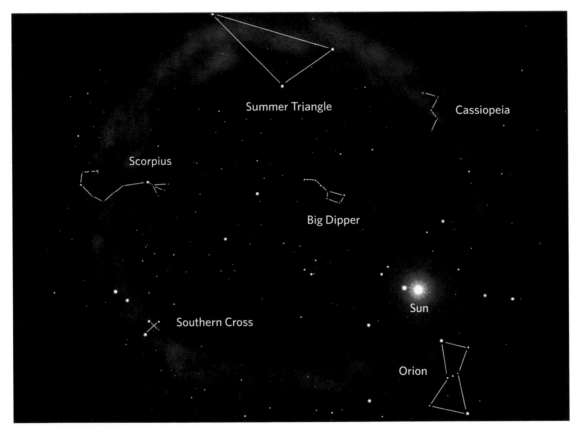

▲ *If you could rocket away from Earth and take a spacewalk, you'd see that the Milky Way makes a complete circle around the sky. That's because the solar system lies in the plane of the galaxy; when we look through that plane, stars stack up in every direction to form a band of milky starlight. Source: Stellarium*

Best times for summer Milky Way viewing (all times are local):

- June—midnight to 3 a.m.
- July—11 p.m. to 2 a.m.
- August—10 p.m. to midnight
- September—9 p.m. to 11 p.m.

List of New Moon dates:

- **2018**: January 16/February 15/June 13/July 12/August 11/September 9/November 7/December 7
- **2019**: January 5/February 4/June 3/July 2/July 31/August 30/September 28/November 26
- **2020**: January 24/February 23/June 21/July 20/August 18/September 17/November 14/December 14

(For future years, consult the resources below.)

RESOURCES

- Light pollution map:
  lightpollutionmap.info

- Moon phases calendar:
  moonconnection.com/moon_phases_calendar.phtml

- About the Milky Way Galaxy:
  imagine.gsfc.nasa.gov/science/objects/milkyway1.html

- Get a real sense for the scale of the universe—interactive:
  scaleofuniverse.com

5

Winter Milky Way

I thought long about whether to include the summer Milky Way's alter ego in this book and decided it would be an oversight not to do so. After all, the run of the galaxy we see in June, July and August, while absolutely magnificent, is only half the story.

The Milky Way's lonelier half stretches like a long, wispy scarf from Cassiopeia, high in the northern sky, across Perseus and Auriga then down through Gemini, just touching Orion at Betelgeuse before widening and brightening near Sirius in Canis Major, pushing past Puppis the Stern and then blending into the haze at the southern horizon.

While a fair number of people have seen or heard of the summer Milky Way, few are acquainted with the fainter winter half. And yet, it's not hard to see. Anywhere you'd easily spy summer's star clouds works for winter, too. From my home on the outskirts of an overly lit, medium-sized city, I still see it rise with Orion and Gemini in the eastern sky every December. This section of the galaxy is thinner in places and not split down the middle by clouds of interstellar dust like the summertime half, but it otherwise looks the same.

▲ *Although it lacks the punch of the summertime Milky Way, the winter half is easily visible as a smoky band from Cassiopeia through Canis Major in a dark sky. Credit: Bob King*

Facing southeast around 9 p.m. in late December

If you look up to where the constellations Auriga and Taurus touch, your gaze takes you to the galaxy's anti-center, the point directly opposite the center of the galaxy as seen by an observer on Earth. Look here and you face the vastness of intergalactic space, a virtually starless gulf that separates our galaxy from the next. Now, close your eyes and picture the billowy star clouds and core of the galaxy at your back, the galactic frontier ahead, with you standing right in the middle of it all. Feel a little giddy? I know I have. Try this on a winter night for a perspective-blowing experience.

As you follow the milky, star-studded band south to Sirius and Canis Major, the Greater Dog, it widens with promise but soon merges with the horizon haze. Skywatchers in the tropics can continue to follow it past Puppis through Carina and Crux, Circinus and Norma until it re-emerges as the summer Milky Way in Scorpius. The whole of the Milky Way wraps around us like a great wreath, a complete circle encompassing both hemispheres.

We see the galaxy's billions of suns all around us because the solar system is neither above nor below the galaxy's flattened disk but embedded *inside* it. When we look through that disk, rather than above or below it, billions of stars stack up across tens of thousands of light-years to form the starry belt of the Milky Way. Most of them are distant, so their light blends into a luminous fog. That's why binoculars and telescopes come in handy. They gather and magnify the light of the distant stars, showing us millions upon millions of them instead of the roughly 5,000 we see with the naked eye on a dark night.

▲   *The Milky Way billows past several of winter's brightest stars and constellations. Source: Stellarium*

So why does the winter half look anemic? The galaxy measures about 100,000 light-years across, with the solar system 26,000 light-years or halfway between the center and edge. In summer, when night falls, we face toward the interior of the galaxy with its massive, star-rich bulge. In winter, we look directly opposite the bulge toward the outskirts of the Milky Way, where stars thin out until giving way to intergalactic space.

How to see the winter Milky Way

If you're now champing at the bit for a more fulfilling Milky Way experience, plan an outing on a moonless winter night. Dress warmly and get out of town to a dark sky. Winter makes finding a spot to park and observe trickier because favorite road pullouts and parking areas can be covered in deep snow. You may have to opt for a gravel road with an open view to the south. I've been there many times. Plan your trip anytime from about three days past full moon up to three to four days past new moon, a two-week window each month. Take a friend along to keep you company and share the experience.

The best evening views will be from mid-December through mid-March, but you can also see the Milky Way to good advantage and in warmer weather if you get up about two hours before dawn in October and November. Here are suggested evening times:

- November—1 a.m. to 4 a.m.
- December—10 p.m. to 2 a.m.
- January—8 p.m. to midnight
- February—7 p.m. to 10 p.m.

No special equipment other than warm clothing and chemical or electric hand warmers is needed. It never hurts to bring along a pair of binoculars to pick out some of the star clusters and nebulae that inhabit winter's fuzzy scarf.

RESOURCES

- Moon phases/lunar calendar. Phase times are shown for your time zone and account for Daylight Saving Time: timeanddate.com/moon/phases/.

- Stellarium or a phone app like Sky Chart described on page 10 will help you identify constellations touched by the winter Milky Way.

Andromeda Galaxy

If all you know is the Milky Way Galaxy, then you simply *must* meet Andromeda. We're fortunate to live in a time when we have some understanding of the size of the universe as well as the true nature of many of its inhabitants. Only a hundred years ago, astronomers were debating whether galaxies were dusty newborn stars residing in the Milky Way or freewheeling galaxies like our own but millions of light-years away. Even the processes by which stars made their own heat and light eluded the best minds of the time.

That's all changed. Hydrogen fusion is the coal of the stars, and galaxies of every shape and size reach to the ends of the visible universe. Although most galaxies require a telescope to see, a few are visible with the naked eye, notably the two Magellanic Clouds, satellites of the Milky Way, and the Andromeda Galaxy. We'll meet the Clouds in a later entry, so let's get to know Andromeda, starting with a couple superlatives. It's 2.5 million light-years away and the farthest

▲ *At 2.5 million light-years away, Andromeda is the closest large galaxy to the Milky Way. Like our galaxy, it's also a spiral but twice as big and composed of a trillion stars. Credit: David Fogel*

thing most of humanity will ever see without optical aid. While it resembles our own galaxy with a flattened disk, a bulge in the middle and spiral arms, it's more than twice as big—220,000 light-years across—and contains approximately one *trillion* stars.

Still not impressed? Andromeda is speeding in our direction at about 250,000 mph (402,000 kmh) and will collide with the Milky Way in four billion years. You needn't lose any sleep over the thought, partly because that's a long time from now, but also because even when galaxies merge, the chance of star hitting star is incredibly remote. But it's a certainty that each galaxy's massive interstellar clouds of gas and dust will collide, compress and give birth to a sparkling array of new stars and star clusters in an eye-popping, galactic fireworks display.

On the other hand, there's also a 50/50 chance the solar system would be flung far from the core of the newly formed super-galaxy or ejected entirely from the system (12 percent chance). In either case, no ill effects would be felt on Earth. That's assuming Earth will be around at that late date. A billion years from now, the Sun will shine brighter than it does today, raising the average surface temperature of the planet to 116°F (47°C). Compare that to today's average of 61°F (16°C).

Then it gets hotter. In three billion years, the average temperature will reach 300°F (149°C). Bacteria might be able to survive the ferocious heat, but little else will. When the two galaxies become one, simple organisms may still find niche environments underground or atop the tallest mountains, but the game of life will likely come to an end 7.5 billion years hence, when the Sun balloons into a red giant star and engulfs the entire planet in its fiery girth. We know this is coming. If humanity can survive itself, we will have long since filled our space tugs with seeds and souvenirs and moved to a planet with a friendlier forecast.

⋀   *If humanity can hold on that long, our distant descendants will witness a most remarkable sight four billion years hence: Andromeda colliding with the Milky Way. Source: NASA*

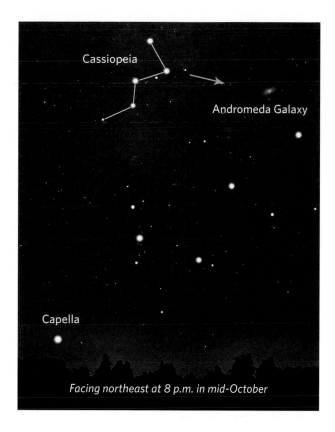

Cassiopeia

Andromeda Galaxy

Capella

Facing northeast at 8 p.m. in mid-October

Andromeda invites big thoughts about time, evolution and a future far different from what even the wisest among us can predict. You can make your own pilgrimage to this galactic oracle on any clear fall or winter evening, when the constellation Andromeda climbs high overhead. Photographs and larger telescopes show the galaxy as a star-spangled cinnamon roll with a butter-hued center and arcs of dark dust silhouetted against the blended light of billions of suns.

With the naked eye, it looks just like a snippet of Milky Way adrift from the mothership. That is to say, it's a dim, fuzzy patch of light about 2° across or not quite the width of your thumb held at arm's length. Using averted vision, the technique of looking *around* a faint object instead of directly at it, you'll notice that the center looks brighter. That's because you're looking at the nucleus, where a great many stars are concentrated.

Have binoculars? They show you even more detail and expand the size of the galaxy at least twofold. A big, oval glow will fill the field of view of a 6-inch (15-cm) telescope at low to medium magnification (40x to 100x); you'll get a clear view of the bright core, the disk and even a hint of those cosmic dust stripes. Larger telescopes uncover the same critters we see closer to home in the Milky Way: open and globular clusters, nebulae and even the occasional nova, a type of stellar explosion. No doubt Andromeda's as rich in planets as our home galaxy. And maybe some of them have eyes looking in our direction wondering who or what might be looking back.

How to see Andromeda

You don't need pitch-black skies to view Andromeda. Outer suburbs, where at least part of the sky is free from serious light pollution, should be fine. Best viewing times, when the galaxy is highest and free from pollution, are listed below. The W-shaped constellation Cassiopeia and the Great Square of Pegasus will point you there. **Culmination** refers to the time when an object is highest in the sky.

▲ *Located in the constellation Andromeda, you can easily find the galaxy by using the "arrow" in the W of Cassiopeia to point you there. Look for a small, fuzzy glow then use binoculars or a small telescope for a better look. Source: Stellarium*

Viewing times

- September—eastern sky, from 9 to 11 p.m. Culminates high in the southern sky around 2 a.m.
- October—eastern sky, 8 to 11 p.m. Culminates around midnight.
- November—eastern sky, 6 to 8 p.m. Culminates around 9 p.m.
- December—western sky, 8 to 11 p.m. Culminates at 7 p.m.
- January—western sky, 7 to 10 p.m. Culminates at 6 p.m.
- February—western sky, 7 to 9 p.m. Culminates before sunset

While you can see the galaxy without optical aid, I strongly recommend binoculars for this and some of our other must-sees. They don't have to be expensive or powerful. Ideal sizes include 7x35, 8x40, 7x50 and 10x50. The first number is the magnification and the second is the diameter of the light-gathering lenses in millimeters.

The binoculars should be easy to hold in your hand, give sharp, in-focus images and have good **eye relief**, anywhere from 14 to 20 mm. Eye relief is how far your eye can be from the eyepiece and still take in the whole field of view. Generally, more is better. If you wear glasses like I do, it's crucial, otherwise, part of the view will be cut off. By the way, most binoculars come with rubber eye cups. If you leave them up, they can compromise eye relief. I always fold them down.

Field of view, the width of the field you see when looking through the binoculars, is also important. Wider fields of view are preferred. They give a picture-window quality to the view, and cover more sky, making it easier to spot what you're looking for. Shoot for a field of view of 5 to 7°.

One final thought about **magnification**. You might think the more power the better. Indeed, you'll find plenty of 15x and 25x binoculars for sale. But extra magnification has drawbacks. First, 15x binoculars are much harder to keep steady in your hands. Every beat of the heart will jostle the view. To effectively use them, they should be fixed to a tripod. Second, both field of view and eye relief generally shrink the more the magnification is increased.

If you have the money, consider a pair of image-stabilized binoculars. These specially designed instruments take the shake out of viewing by using a battery-operated stabilization mechanism. Press the button, wait two seconds and the shake disappears even as you move the binoculars from target to target. The detail and fainter stars visible may convince you they're worth the price!

Check the resources section on page 216 at the end of the book for specific model suggestions. No matter what you buy, try out a variety of binoculars at an outdoor recreation store before making a purchase.

RESOURCES

- Stellarium or a phone app like Sky Chart described on page 10.
- The SkyLive—know where to look for the galaxy anytime from anywhere: theskylive.com
- More on the mash-up of the Milky Way and Andromeda: science.nasa.gov/science-news/science-at-nasa/2012/31may_andromeda
- Best binoculars and binocular reviews: bestbinocularsreviews.com

Ursa Major the Great Bear

Some of our must-sees require travel or special circumstances, such as an eclipse, but not Ursa Major the Great Bear. I included this constellation because it's visible much of the year and bears a good resemblance to its namesake, something that can't be said about some of the other 87 sky figures.

You're probably already familiar with the seven bright stars that outline the Big Dipper, an eye-catching asterism that wheels around the Polestar in the northern sky. Many of us picture it as a ladle or long-handled pot, but it's the Plough (plow) in Britain. Long before there was an America, most Europeans saw it as a wagon or chariot.

⌃   *Summer and early fall are the easiest times to use the Big Dipper to trace out the rest of the Great Bear figure. Find a place with a reasonably dark sky and a good view to the northwest. Credit: Bob King*

31

You might wonder then how it came to be called a bear. There's a clue in its name—the constellation is an Urs*a*, or she-bear. Some scholars trace the constellation's origin to Callisto, a nymph from ancient Greek mythology, who was transformed into a bear by Hera, the wife of Zeus, as punishment for having an affair with her husband.

Later, while foraging in the woods, Callisto spotted her son, Arcas, who was out hunting. Forgetting she was a bear, Callisto ran toward him, but Arcas, seeing an easy kill, drew his bow and fired an arrow. Before it struck, Zeus sent a great wind to carry them both up to the heavens, transforming Callisto into Ursa Major and Arcas into the neighboring constellation, Boötes. Both travel in tandem around the Polestar to this day.

Even further back in time, Native Americans, including the Iroquois and Great Lakes Indians, saw a different sort of bear here, based on their own traditions. The three stars in the handle were three brave warriors in pursuit of a bear, outlined by the four stars in the Dipper's bowl. Mizar, the star in the bend of the handle, has a little companion star named Alcor, which represented the pot for cooking up their catch. When the hunters finally catch up with the bear, they spear it; the blood that drips from the wound is said to turn the leaves red every fall.

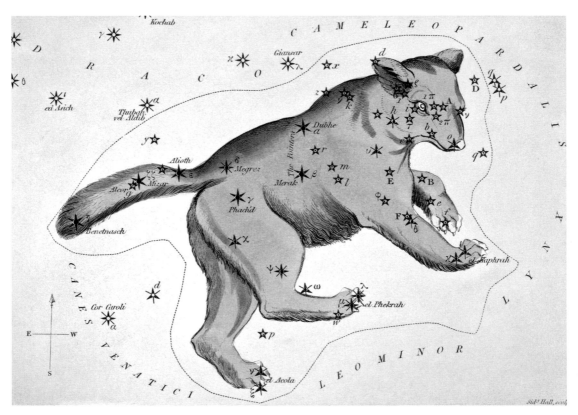

▲   *The Big Dipper makes up only the brightest part of the much bigger figure of the Great Bear. Source: Sidney Hall–Urania's Mirror*

For all its familiarity, the Big Dipper isn't a true constellation but an **asterism**, a bright pattern of stars within a constellation—or shared by multiple constellations—that's easy to recognize. Asterisms are helpful waypoints to navigate to a constellation's fainter stars. The Big Dipper makes it easy to find the fainter stars that outline Ursa Major.

While the Great Bear is visible nearly year-round from midnorthern latitudes, it looks most impressive from mid-to-late summer, when the constellation pads toward the northwestern horizon like its namesake, ambling through the forest looking for a good hibernation location. In winter, the bear climbs up the northeastern sky standing on its tail like some cartoon circus bear; in spring, it's high overhead and upside-down; and in winter, paws and legs get cut off by the northern horizon.

Summer's the season to check this celestial bear off your list. If nothing else, do it for Callisto, who as it turns out was falsely accused, forced into the tryst against her will by the devious Zeus.

How to see the Great Bear

Find a location with a reasonably dark sky and face northwest at nightfall during one of the times listed below. Halfway up in the northwestern sky, you'll see the familiar seven stars of the Big Dipper. Picture the three in the handle as the bear's tail and the two across the top of the bowl as the animal's rump and back. Extend a line from the bowl to the right (west) to pick up the several stars that outline the head and tip of the nose then double back to the back of the bowl. Now, follow the arc of stars just below and to the left downward to trace out a leg and paw that ends in two claws. Return to the right side of the bowl and extend a line directly below to find a skinny second leg and paw tipped by claws. Arc back up to the bottom of the head and tip of the nose to complete the figure.

Not a bad likeness, wouldn't you say?

Although Ursa Major is visible for much of the year, it looks most bear-like in the northwestern sky at these times:

- May: 2 to 4 a.m.
- June: 10 p.m. to 1 a.m.
- July: 10 to 11 p.m.
- August: 9 to 10 p.m.

Total Lunar Eclipse

Chances are you've seen a partial lunar eclipse, when the Moon slips partway into the shadow cast by the Earth. You might be carrying groceries to the car, look up and see an odd-looking bite taken out of the Moon. While an enjoyable sight on its own, a partial eclipse is more of an appetizer. For the "full meal" with all the trimmings, only a total eclipse will do.

In a typical year, there are four solar and lunar eclipses. They can be total, partial or a mix of both. Some years have five, six and, rarely, seven eclipses. The last seven-event year happened in 1982 with four partial eclipses of the Sun and three totals of the Moon; the next eclipse-a-copia occurs in 2038 with three of the Sun and four of the Moon.

A total lunar eclipse occurs when the Earth lies exactly between the Sun and the Moon. If we could zoom out into space and look back at the scene, we'd see the Sun, Earth and Moon neatly lined up in that order. Lunar eclipses only occur at full moons because only then does the Moon sit opposite the Sun on the shadowed side of Earth.

▲ *The Moon glows orange-red over Duluth, Minnesota during the January 31, 2018 total lunar eclipse. Lunar eclipses can occur up to three times a year (but as few as zero), when the Full Moon moves into the shadow cast by the Earth. Credit: Bob King*

Just like you and I do on a sunny day, the Earth casts a shadow. If your shadow happens to fall over a patch of flowers in a garden, you might describe the blooms as being in total eclipse. That description would raise an eyebrow or two, but you'd be correct. Keep walking and the eclipse would end as your shadow moved past.

The next time you're out for a stroll, take a closer look at your shadow, and you'll discover that the inside part is darker and the edge lighter and fuzzy. The darker part is called the **umbra**, the lighter border, the **penumbra**. Inside the umbra, the Sun is completely hidden from view, while in the penumbra, it's only partially blocked. Light from the Sun filters in to brighten the edge.

You'll also notice that the penumbral outline of your shadow has a soft, fuzzy edge, not a crisp, sharp border. That's because the Sun is an **extended object**, not a point of light. Light from one side of the Sun's disk can filter into areas that are shadowed from the light coming off the other edge and vice versa. If the Sun were a brilliant, laser-like point, shadows would be sharp and uniformly dark without fuzzy penumbras.

Like our own shadow, Earth's shadow also has a fuzzy, blurry edge (the penumbra) and a dark, inner umbra. The umbral shadow extends about 870,000 miles (1.4 million km) into space. At the Moon's distance, it's about 2.6 times the Moon's diameter or about 1.3° wide. To appreciate how small that is, raise your index finger to the sky. The bit of blue your fingertip covers is the width of Earth's shadow where the Moon orbits. The penumbral shadow, which encircles the umbra, spans 2.9° or about the width of your index finger paired with your middle finger.

For a total eclipse to occur, the Moon must hit that small target. It would be simple if the Sun, Earth and Moon were lined up in the same plane in space like runners on a track, but the Moon's orbit is tipped 5.1°. During most lunar cycles, it passes a few degrees above or below the shadow, completely missing it. Instead of an eclipse, we see a gorgeous full moon.

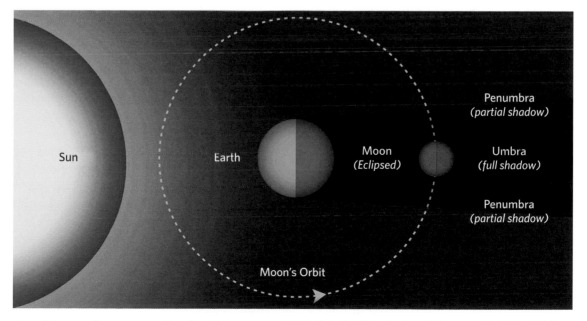

▲ *A lunar eclipse occurs when the Moon passes directly into Earth's shadow. This can only happen when the Sun, Earth and Moon are in line and the Moon is full. Source: Starry Night*

Other times, it's closer to its target but not dead-on. For instance, the Moon might only pass through the penumbra for a lightly shaded **penumbral eclipse**. Or it might put just a "toe" in the deeps, dipping partway into the shadow in a partial eclipse. But when it treads all the way into the umbra, not only do we witness a near-perfect alignment of three celestial bodies, but the fully eclipsed Moon looks strangely unfamiliar. Even otherworldly.

Bereft of bright sunlight, our satellite appears faint and far away. Without moonlight to scrub away the night, darkness returns with a surprising suddenness. Stars wink back on. The transition brings a hush to the scene and a mood of quiet reverence.

Despite Earth's shady ways, the Moon doesn't completely disappear. If the planet had no atmosphere, it would. But the air acts like a prism and refracts a smidge of sunlight into the shadow to color the Moon burnt orange, smoky red, copper or even muddy brown.

Here's how it works. Sunlight streaming through space in the eclipsed Moon's direction gets refracted, or bent downward, into the shadow cone when it encounters Earth's atmosphere in exactly the same way it's bent when it travels from air through a glass prism. Because it passes through the atmosphere at a very shallow angle—the same angle it does at sunrise and sunset—the air filters out all the blues and greens from sunlight, leaving only the reds and oranges to penetrate the shadow.

As incredible as a lunar eclipse appears from Earth, it would be even more amazing to stand on the Moon and look back at the Earth eclipsing the Sun! Both happen simultaneously during every total lunar eclipse. A lunar astronaut would see our planet rimmed in 360° of sunset-red air and sihouetted against the sun's outer atmosphere called the **corona**.

The color of the fully eclipsed Moon varies due to many factors including how deeply it travels into the umbra and the state of Earth's atmosphere. It's particularly sensitive to volcanic dust suspended in the atmosphere in the aftermath of a major eruption. One of the darkest eclipses in recent memory occurred on December 30, 1982 due to the effects of the lingering ash cloud from the El Chichon eruption in Mexico in March that year.

Orbiting at an average speed of 2,288 mph (3,683 kph), the Moon takes about an hour to cross the umbra. If you add in the partial phases before and after, an eclipse lasts about four hours. Many observers ignore the penumbral phase when the Moon crosses the pale outer shadow coming and going. But if you include the time spent in the penumbra, a total lunar eclipse can last around six hours! What's more, unlike a total solar eclipse, a lunar eclipse is visible wherever the Moon is up in the sky, meaning that half the planet can see it.

As the Moon slides into and out of eclipse, you can fully enjoy the color and shape changes without optical aid. But I recommend at least a few peeks through binoculars to better discern the color gradations across the Moon's face. You'll also have the rare chance to see the Moon suspended in apparent 3D among the twinkling stars. Recall that the glare of a normal full moon washes away any stars near it. Seeing both simultaneously during totality will send you into orbit.

A small telescope is handy too. Nothing fancy. You'll better appreciate the fuzziness of Earth's shadow as you look through the eyepiece. You might also catch sight of a turquoise edge to the shadow caused by the ozone layer that extends from 10 to 30 miles (16 to 48 km) overhead. Ozone absorbs orange and red light, lending a bluish tint to the encroaching shade.

Be sure not to miss my personal favorite part of a total eclipse, the few minutes before and after totality, when the narrow rim of the Moon is still bathed in partial sunlight while the rest of the globe glows red. I swear it looks just like Mars topped by one of its polar ice caps—only big and close!

Consider using your mobile phone camera to record the progress of the eclipse. It's easy. Center the Moon in the eyepiece of a telescope, then hold the camera lens directly over the eyepiece until the Moon fills up the screen and snap away. Phones take surprisingly good photos, and you can instantly post it for your friends.

How to see a lunar eclipse

First, use the table below to identify an eclipse happening where you live, then as the time approaches, stay in touch with the weather forecast for your region. If it's clear, you're all set. If not, then use a satellite weather map or check a regional forecast to find clear skies. You'll find helpful websites below.

Pay attention to eclipse times to plan your outing. Do you want to see the entire event, just totality or a little bit of both partial and total?

The list below includes all total lunar eclipses through the year 2029. Dates are centered on Eastern Standard Time for eclipses visible in the Americas and Greenwich Time when visible in the eastern hemisphere. Eclipses that are visible at least somewhere in the United States are shown in bold:

- July 27, 2018—South America, Europe, Africa, Asia, Australia
- **January 20, 2019**—Americas, Africa, Europe, central Pacific
- **May 26, 2021**—western North America, Australia, East Asia, Pacific
- **May 15, 2022**—Americas (except Alaska), Africa, Europe
- **November 8, 2022**—Most of the Americas, Asia, Australia, Pacific
- **March 14, 2025**—Americas, western Europe, western Africa, Pacific
- September 7, 2025—Europe, Africa, Asia, Australia
- **March 3, 2026**—Most of the Americas, East Asia, Australia, Pacific
- December 31, 2028—Europe, Africa, Asia, Australia, Pacific
- **June 25, 2029**—Americas (except Alaska), Europe, Africa, Middle East
- **December 20, 2029**—eastern Americas, Africa, Europe, western Asia

RESOURCES

- NASA Eclipse website: Full details of past and future total, partial and penumbral lunar eclipses with times and visibility maps. Highly recommended! eclipse.gsfc.nasa.gov/lunar.html

- National Weather Service forecasts: Just enter your city name. weather.gov

- GOES East current weather satellite photo: weather.msfc.nasa.gov/GOES/goeseastconus.html

- GOES West for the western United States: weather.msfc.nasa.gov/GOES/goeswestpacus.html

- ClearDarkSky site: Find your town and get an hour-by-hour detailed cloud forecast at cleardarksky.com

- Interactive U.S. cloud cover forecast: weatherstreet.com/states/u-s-cloud-cover-forecast.htm

9

Total Solar Eclipse

No astronomical sight stirs more emotions than a total solar eclipse. During the precious minutes the Moon covers the Sun, some weep and others shout for joy. Logic fails at totality, and so it should. On the surface, an eclipse is a simple thing. The Moon comes directly between the Earth and the Sun like someone who walks in at the last moment to block the view of your favorite band at a concert. For a few minutes, the Sun disappears, replaced by the black silhouette of the Moon against the solar **corona**.

This is straightforward stuff, but let's take a closer look. First, all solar eclipses happen at the new moon phase. That's when the Moon lines up in the direction of the Sun. We normally can't see a new moon because it's completely washed out by bright daylight. Only during a partial or total solar eclipse when the Moon precisely (or nearly so) lines up with the Sun does it become visible (but of course only in silhouette).

Since a new moon like a full moon happens about once a month, why don't we have total eclipses a dozen times a year? We would if the Moon's orbit weren't tilted 5.1° with respect to Earth's orbit. That tilt means that most new moons pass either north or south of the Sun and completely miss it. When the Moon is almost but not quite in line, we see a partial eclipse. But only when the line-up is exact, which occurs on average once every eighteen months, does someplace on Earth experience a total solar eclipse.

Total eclipses are uncommon enough, but the chances of any particular location getting one are downright rare. I've seen different estimates, but it's typically a 300-year wait before an eclipse simply arrives at your doorstep. For instance, the last total eclipse seen in Minneapolis, Minnesota, occurred on June 30, 1954, and the next will be October 14, 2099.

When the Moon passes in front of the Sun, it casts a shadow on Earth, called **the path of totality**, that can run for thousands of miles in length but typically is only 60 to 100 miles (100 to 160 km) wide. On maps, the shadow path looks like a long noodle. This is very different from a lunar eclipse, where half of the planet—basically wherever it's nighttime—can see the totally eclipsed Moon. You can only see the totally eclipsed Sun from somewhere inside that narrow ribbon of darkness defined by the shadow of the Moon as it moves across the Earth.

Because the Moon continuously orbits about Earth, its shadow moves along the ground, traveling from west to east, the same direction the Moon travels around the Earth. The Moon's average orbital speed is 2,288 mph (3,683 kmh), so you might think the shadow moves that fast, but it doesn't because the Earth is rotating in the same west-to-east direction with a speed that varies according to latitude. At mid-latitudes, such as in the United States, Europe and China, the planet whips around at about 750 mph (466 kmh), so we have to subtract that number from the Moon's speed.

Other factors play a part, too. The shadow is projected onto a spherical Earth, not a flat surface. It first touches the planet at a very oblique angle, moving extremely fast, but then slows down near the middle of the totality path and picks up speed again as it departs Earth. Factoring it all in, the shadow still hurries along at over 1,000 mph (1,600 kmh) on average. During the August 21, 2017 solar eclipse, witnessed by tens of millions of people across the lower 48 states, the shadow took only about 90 minutes to race from the Oregon coast to the South Carolina shoreline. No wonder totality at any given location only lasts a few minutes at most—the Moon's shadow moves swiftly!

Now you can begin to understand why most folks never get to see a total eclipse. Many of those eclipse noodles cross the middle of the Pacific, the frozen shores of Antarctica, central Siberia and lots of other places that cost big bucks to get to. But even if you pony up the cash and go, there's no guarantee the weather will cooperate. My first two solar eclipses—both about 1,500 miles (2,400 km) from my home—were clouded out. I did better on the next four but only two were cloud-and-worry-free.

▲   *A total solar eclipse is one of the grandest, most emotional experiences in all of astronomy. During totality, the Moon covers the Sun completely, revealing its luminous atmosphere called the corona, pictured here during the August 21, 2017 total eclipse. Credit: John Chumack*

Eclipses are addictive. Once you've seen one totality, you crave more. Never mind the miles. Get in the car, hop on an airplane, hook up with a tour group—whatever it takes to witness one of nature's most extraordinary sights.

Prior to totality, you can watch the Moon take its first "nibble" of the Sun's disk. When 80 percent of the Sun is covered about an hour later, the change in daylight becomes evident. Sunlight is weaker. At 90 percent, the daylight change becomes dramatic as if seeing the world through sunglasses. The temperature drops as the Moon chokes off the Sun's light *and* heat. Temperatures can drop more than 10 degrees.

This slipping away of the light has a profound emotional impact on eclipse watchers just as it must have had on our distant ancestors who had no scientific explanation for the event. We can't help but feel a twinge of fear, or even a sense of impending apocalypse, at seeing the Sun about to disappear in the middle of the day. It's *not* normal.

In the few minutes remaining before totality, you'll notice how sharp shadows become compared to the fuzzy ones that follow us every other day. That's because the Sun is a narrow crescent and closer to being a point source than a disk. Point sources produce sharp shadows.

Moments before totality, watch the western sky for a large, dark presence headed in your direction; that's the approaching shadow of the Moon. With a suddenness that will almost knock you over, the last beads of sunlight pop and twinkle along the Moon's advancing edge and then it happens—the lid covers the pot. The Sun vanishes, and in its place, the black disk of the new moon appears, ensconced in the unearthly, amoeba-like glow of the corona. The sight is unlike anything else you'll ever see.

If you can wrest your eyes from the scene, look around to experience the twilight-like darkness and the bright stars and planets now visible in the middle of the day. Venus is almost always present and easy to spot.

▲   *In this August 21, 2017 eclipse sequence, both the observing party and a mountain range to the west are in sunlight before total eclipse (top). The Moon's shadow, which approaches from the west, first covers the mountain range (center) before engulfing the eclipse watchers seconds later (bottom). Credit: Tom Nelson*

Listen for crickets, owls and look for other changes in animal behavior. During the 2017 eclipse, my younger daughter, Maria, alerted us to the crickets, while my older one, Katherine, keenly observed that the two dogs in our company, which had been resting and panting in the shade earlier, jumped up and playfully chased each other during totality.

Totality is a wildly beautiful thing, provoking shouts of joy and even applause. Others sob or stand quietly, absorbing a sight for which they have no words. Many of us take photos. One tip on photos: snap a few as souvenirs, but spend most of the time taking in the scene. This may be the only eclipse you'll ever see.

Much of the corona's structure is visible with the naked eye, but I encourage you to spend a little time looking through binoculars or a small telescope to appreciate the fantastic detail in the streamers that extend away from the Sun and the neatly aligned plumes crowning its north and south polar regions.

The corona isn't an atmosphere as we normally think of it, but sunlight reflecting off electrons, the tiny particles in electricity, and some dust as well. The electrons trace the magnetic fields of the Sun in exactly the same way iron filings sprinkled around a magnet reveal the curlicues of otherwise invisible magnetic energy. Some of the Sun's electrons are cast away into space in daily zephyrs called the **solar wind**. When they reach Earth, they can sometimes hook into the planet's magnetic field and spark auroras. Before the Sun returns, be sure to look for ruby-red "flames" poking out from the moon's edge called **prominences**. Towering tens of thousands of miles high, they're made of incandescent hydrogen gas.

Would that we could command the Moon to stop, but it moves on. A few minutes after snuffing out the Sun, daylight returns with a shock as fresh beads of sunlight pop into view along the Moon's western edge. This "first light" is sunlight shining through low spots between craters at the lunar limb and called **Baily's beads**. The brightest of them, seemingly attached to the bright corona surrounding the Moon, creates a momentary "diamond ring" effect.

If you've ever wondered if the moons around other planets in our solar system cause eclipses, they do. Mars's tiny moons, Phobos and Deimos, eclipse the Sun, but they're so small, only partial eclipses occur. The outer planets have lots of moons and lots of eclipses but one big problem: no solid surface from which to watch. They're all atmosphere. To see Jupiter's moon Io eclipse the Sun, you'd have to view it from the window of a spaceship.

Sometimes, the Sun–Moon–Earth line-up is precise but the Moon happens to be near apogee, its most distant point from Earth, and appears smaller than usual. Instead of a full eclipse, observers witness an **annular** or "ring" eclipse. The day darkens but nothing like totality.

I've seen two annular eclipses. Besides a nice display of Baily's beads, it's a thrill to watch the Moon move in real time across the Sun's face especially when observed through a telescope.

How to see a total solar eclipse

You can look up any eclipse, when it happens, and the duration and location of the path of totality using the NASA Eclipse website listed below. That way, you can find one that's suitable for your income and time frame, whether you're driving, flying or taking a cruise ship. Eclipse tours are extremely popular and easy to look up online. Among others, *Sky & Telescope* magazine, the Planetary Society and TravelQuest International all offer tours along with eclipse glasses, pleasant places to visit along the way and observing sites for participants.

The only must-have equipment for a total solar eclipse is a safe viewing filter. Approved eclipse glasses or a #14 glass welder's filter are perfect for naked eye viewing. You'll need them for all the partial phases, but remove them during totality, when it's perfectly safe to look.

For telescopes, you'll need a filter to cover the front end of the tube. I also suggest using binoculars to study the corona during totality. A GoPro or similar wide-field video camera that you can switch on and forget about makes a nice addition.

RESOURCES

- NASA Eclipse website/Eclipses from 2001 to 2100: eclipses.gsfc.nasa.gov/SEcat5/SE2001-2100.html

- Rainbow Symphony—for approved eclipse filters and glasses: rainbowsymphony.com/

- *Sky & Telescope* eclipse tours: skyandtelescope.com

- TravelQuest International tours: travelquesttours.com/all-tours/

10

Awesome Aurora

What astronomy news gets a lot of hits on the Internet? I can tell you. Close-approaching asteroids, giant solar flares, an amazing new Hubble discovery, bright comets . . . and the aurora borealis.

If you live in the southern half of the United States and points south of the U.S., you're probably a stranger to the aurora unless you've traveled to the North for a visit or vacation. With luck, you arrived at the same time as a parcel of energized particles from the Sun smacked into Earth's magnetic domain and sent it into a frenzy. If you'd rather not rely on luck, there are more deliberate ways of finagling a sight of the northern lights.

Take a tour! Lots of outfits now offer special aurora packages to places such as Iceland, Alaska and Greenland, where the aurora puts on a show as often as the nightly news. We'll explore some of these places in just a bit.

In the aurora, nature combines a magnificent light show and thrilling emotional cocktail made of two parts wonder and one part fear—the safe kind. A grand display can turn a respectable citizen into a sleepless zombie, lurching to work the next morning after staying up all night, unable to carry on a proper conversation outside of "Wow." "Awesome." "Unbelievable."

Auroras are transformational—raw displays of nature's power writ across the sky in flaming curtains, rippling waves and jabbing fists of light. Shape-shifting moment to moment, each display

▲ *A raging, all-sky aurora makes for an unforgettable experience. Credit: Bob King*

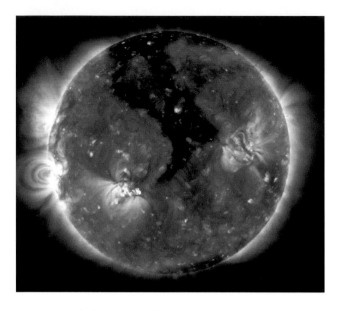

is as unique as a snowflake, and yet they all follow a familiar rhythm of waxing and waning intensity. All that energy crashing over our heads can make us feel powerless, a refreshing emotion in a world where some hold on too tightly.

Nature, ever inspired by necessity and a willingness to use any and all spare parts available, creates the spectacle with the tiniest of things—subatomic particles. We're talking protons and electrons, but mostly electrons, the same stuff that powers your toaster. The Sun is made of mostly hydrogen—ordinarily a single proton orbited by a single electron—but extreme temperatures strip away the electrons to create a mix of freely moving protons and electrons, what physicists call a **plasma**.

Because plasma is made of electrically charged particles (electrons are negatively charged, protons positive), it conducts electricity. Moving charges also create looping magnetic fields just like the ones you see when you sprinkle bits of iron around a magnet.

Disturbances such as a solar flare, a coronal mass ejection or a breach in the solar atmosphere called a **coronal hole**, can launch billions of tons of solar plasma into space at more than a million miles per hour. Most of the time, the material slides past Earth's protective magnetic field like water off a duck's back and we're no worse for wear. But if the south end of the plasma's magnetic field happens to brush over the north end of Earth's field, opposite poles attract, and the two "snap" together like the north and south poles of a pair of magnets. Clack!

Earth's magnetic field lines guide the plasma toward the polar regions as it accelerates to one-tenth the speed of light. When the speedy electrons slam into the atmosphere, they energize oxygen and nitrogen atoms. But only momentarily. The atoms return to their relaxed states by shooting out spitballs of colored light called **photons**. Oxygen releases green and red photons and is the primary source of the most common aurora colors. Nitrogen spits blue and red.

We can "see" parts of Earth's magnetic field in the multiple, parallel rays common in auroral displays caused by electrons cascading down bunched-together field lines like so many firemen on fire poles. Scientists call an aurora event a **geomagnetic storm** because it's a disturbance in Earth's (the "geo" port) magnetic field.

Auroras in the southern hemisphere—called the **aurora australis**—happen the same time as auroras in the north. Northern lights can occur in any season, but due to the Earth–Sun orientation at the equinoxes, they're more common at those times. Auroras can also happen any time of night. Some of the finest displays I've witnessed were already well under way at dusk. More often, the

▲   *Visible in far ultraviolet and X-ray light, coronal holes are regions where particles from the Sun flow freely into space unconstrained by solar magnetic fields. Source: NASA / SDO*

aurora begins as a glowing arc of pale green light low in the northern sky at nightfall and becomes more intense closer to local midnight.

If you live above latitude 44° north, make a habit of checking the northern horizon for these arcs. An arc might last an hour and fade away, but if conditions are right, it could double, brighten and sprout long, parallel streamers called **rays**. Rays can sometimes reach beyond the zenith and into the southern sky like great searchlight beams. When the aurora peaks, it often forms a corona of rays that flash and twist near the top of the sky followed by fast-moving waves or pulses of light shooting up from the north.

How to see the aurora

The best way to anticipate the aurora is to regularly check the aurora forecast and planetary K-index (Kp index) posted online by the National Oceanic and Atmospheric Administration (NOAA) and updated multiple times a day.

The Kp index is a measure of magnetic activity high in the atmosphere rated from 1 (low) to 9 (high). When the index is low (Kp = 1 to 4), don't expect much for auroras. But when it rises to 5 and you live in the northern United States or southern Canada, there's a reasonable chance you'll see a small to moderate level of auroral activity. At Kp = 6 to 7, the aurora might be seen as far south as the Midwest, while at Kp = 8 to 9, even the southern states will get a view.

Normally, the aurora is anchored over the polar regions as the northern and southern auroral ovals, permanent doughnut-shaped regions of auroral activity about 2,500 miles (4,000 km) across, under which, the Earth rotates. During a geomagnetic storm, the northern oval expands southward into southern Canada, the northern United States, northern Europe and Russia. The stronger the storm and higher the Kp index, the further south the aurora extends and the greater its intensity.

If you don't want to wait for the northern lights to come to you, you can go directly to them by booking one of several tours listed in the resources section on page 217. You'll be visiting a place located under the permanent oval, where auroras are active almost every night of the year. Whatever trip you book, be aware that the weather has to cooperate, too. Make sure you stay at least several nights when there's little to no Moon in the sky.

RESOURCES

- Planetary K-index: swpc.noaa.gov/products/planetary-k-index

- Aurora 30-minute forecast with a regularly updated aurora oval image: swpc.noaa.gov/products/aurora-30-minute-forecast

- Subscribe to all the above services via email: pss.swpc.noaa.gov/RegistrationForm.aspx

- Aurora app for Android: I recommend Aurora forecast at: play.google.com/store/apps/details?id=com.tinacinc.auroraforecast&hl=en

- Aurora app for iPhone: itunes.apple.com/us/app/my-aurora-forecast-northern-lights-borealis/id1073082439?mt=8

11

Southern Cross

We arrived on Grand Cayman Island for our honeymoon on a warm, humid night in mid-April, stepping off the plane onto the tarmac, I turned around to face south and there it was—the Southern Cross. Back home, the Cross never climbs above the horizon. Seeing it here made the 2,000-mile flight as much as an astronomical journey as a terrestrial one.

The Southern Cross is the quintessential constellation of the southern skies. If you ask a northerner to name something in the Southern sky, you're not likely to hear Dorado the Goldfish or Musca the Fly. Nope, it's the Southern Cross or more formally, Crux. There are good reasons for this. While the Cross is the smallest of the 88 constellations—you can easily cover the figure with three fingers held at arm's length—it contains three of the sky's 25 brightest stars: Acrux, Becrux and Gacrux. A line extended through the long axis of the cross to the south points very close to the southern polestar, Sigma Octantis, much the same way the Pointer Stars in the Big Dipper guide us to the North Star.

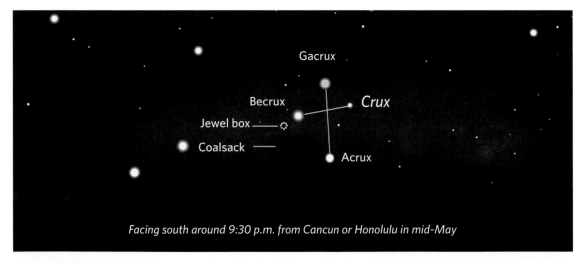

Facing south around 9:30 p.m. from Cancun or Honolulu in mid-May

▲ *The Southern Cross lies below the horizon for all of Canada and much of the U.S., but it can be seen very low in the southern sky in spring and summer from Cancun, Mexico, Hawaii and Key West. The Jewel Box star cluster adds extra sparkle—use binoculars. Source: Stellarium*

Southern hemisphere residents so identify with the Cross that it's featured on the flags of Australia, New Zealand, Brazil, Papua New Guinea and Samoa. For northerners, Crux is the key to learning the constellations of the southern spring sky. Find it and you'll be on your way to nearby Centaurus the Centaur, Lupus the Wolf and Vela the Sails.

Set smack dab in the middle of the southern Milky Way, the Southern Cross region is rich in astronomical goodies. From a dark sky, you'll immediately be struck by a big hole to the left or southeast of the Crux that looks as if someone poured ink on the stars. This is the Coalsack Nebula, an enormous cloud of interstellar dust located 600 light-years from Earth and one of the most prominent of the "dark nebulae" visible to the naked eye. At this very moment, denser clumps of dust in the Coalsack are collapsing to form young protostars. Millions of years from now, our descendants will look here and see that these "lumps of coal" have ignited to form a new generation of stars and star clusters.

Serendipitously, the youthful star cluster nicknamed "The Jewel Box" and unrelated to the Coalsack glimmers along its northern border just 1° from Becrux. Bright enough to glimpse with the naked eye and binoculars, it's one of the most gorgeous sights you'll ever see in a small telescope. I've viewed it a couple times in my little 3.5-inch (88-mm) refractor on southern vacations, but I'm long overdue for another look.

▲ *The Southern Cross or Crux is seen here along with the dark blotch called the Coalsack, made of stardust. Credit: ESO / Yuri Beletsky*

Here's how nineteenth century astronomer John Herschel, who gave the cluster its name, described the sight in *Results of Astronomical Observations Made During the Years 1834, 5, 6, 7 & 8 at the Cape of Good Hope*: " . . . this cluster, though neither a large or a rich one, is yet an extremely brilliant and beautiful object when viewed through an instrument of sufficient aperture to show distinctly the very different color of its constituent stars, which give the effect of a superb piece of fancy jewelry."

You nailed it, John.

How to see the Southern Cross

Live in or near Key West, Florida? Brownsville, Texas? Honolulu? Crux stands very low in the southern sky when the constellation reaches its greatest altitude, or culmination, on spring evenings anywhere south of 26° north latitude. Not sure of your latitude? Use the latitude finder link below.

Culmination occurs in early April around midnight local daylight time; early May around 10 p.m. and early June around 8 p.m. You'll need clean, clear skies and the ability to see down to the southern horizon to capture the cross.

Things get easier the further south of 26° north you live or travel. Personally, I'd book a trip to Cancun (latitude 21°N), Puerto Vallarta (20.6°N), Cozumel (20°N) or Costa Rica (~10°N). After a night of stargazing, you can loll away the day at the beach, eat fish tacos, watch a sunset and then head back out at nightfall for starry desserts.

Aw, what the heck. Just cash it all in and go to Alice Springs, Australia. Located in the heart of that continent 23.7° south of the equator, the night skies are spectacular. The Southern Cross stands more than five fists high at culmination and remains visible for many months of the year, though it's still best in spring, er, fall, since we're talking the *southern* hemisphere, where seasons are reversed.

Use the included map or a star-charting app to help you find Crux, the Coalsack and Jewel Box Cluster whenever they're in view.

RESOURCES

- Latitude finder: latlong.net

- How to find the south celestial pole: oneminuteastronomer.com/3307/finding-south-celestial-pole/

- Online Planetarium—The Sky Live, an excellent, interactive star map of the entire sky from anywhere on Earth. Just select your city and you're set! theskylive.com/planetarium

12

Alpha Centauri

If there's one star we've all heard of in the southern hemisphere sky, it's Alpha Centauri, the brightest or "alpha" star in the constellation Centaurus the Centaur. Its fame comes from proximity—of all the trillions of stars in the universe, Alpha Centauri is the closest star system to the solar system. Notice I said system. We see a single star with the naked eye, but a telescope will show three.

The brighter pair, Alpha Centauri A and B, revolve around each other every 79.9 years and are currently (in 2018) near the minimum separation as seen from Earth. But they're still not so close together that a 3-inch (76-mm) telescope magnifying at 100x can't split them apart. The pair of suns lie 4.37 light-years from Earth. What exactly does that mean? Moving at 186,000 miles per second or 5.8 trillion miles (9.3 trillion km) a year, it takes light 4.37 years to reach us from the star. If we multiply 5.8 trillion x 4.37, the distance to the Alpha Centauri pair comes to a little more than 25 trillion miles. Whew, that's a long way.

The fastest speed attained by a man-made object as of this writing was NASA's Juno spacecraft. When it arrived at Jupiter on July 4, 2016, the planet's gravity briefly accelerated it to 165,000 mph (265,000 kmh) in relation to the planet. Let's say we could launch a rocket from Earth to Alpha Centauri traveling at that speed. How long would it take to get there? About 17,300 years. Yikes.

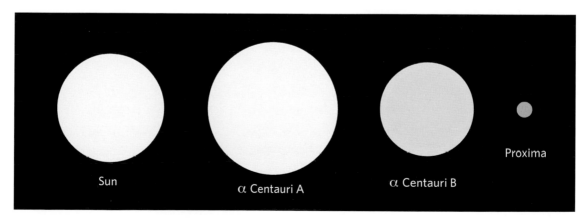

Sun   α Centauri A   α Centauri B   Proxima

▲   *The Alpha Centauri star system is the closest to Earth after the Sun and comprised of three stars: Alpha Centauri A & B and the faint red dwarf, Proxima. Alpha Centauri A is about 25% larger than the Sun. Source: Wikipedia /CC BY-SA 3.0*

Another way to visualize Alpha's distance is to shrink the Sun's size to a 4-inch (102-mm) sphere. At this scale, Jupiter would be a marble 200 feet (60 m) away, Pluto a grain of sand a quarter-mile from the Sun and Alpha Centauri . . . drum roll please . . . a pair of spheres 5 inches (127 mm) and 3.5 inches (88 mm) across 2,000 miles (3,220 km) away! Relatively speaking, space is so empty that matter as we know it hardly exists.

But let's move on to more pleasant thoughts. Truth be told, the bright AB pair isn't actually the *closest* star. That honor goes to Alpha's third companion, a small, faint red dwarf star called Proxima Centauri. Proxima orbits a quarter light-year from the bright pair, putting it a tad closer to our solar system at 4.24 light-years. Unfortunately, eleventh magnitude Proxima is too faint to see with the naked eye or binoculars, but a small 4-inch (100-mm) telescope will show it.

It was thought for a time that an alien planet, called an **exoplanet**, revolved around Alpha Centauri B, but a later analysis confirmed that the "object" was an artifact and not real. Pity. How cool would it have been to look at Alpha Centauri and know it carried a planet in tow? Now for the good news. It turns out that little Proxima has one instead.

Facing south from Cancun, Mexico 10 p.m. June 1

▲   *The bright pair of stars, Alpha and Beta Centauri, is found just to the left or east of the Southern Cross. Proxima, currently the closest star, is faint and requires a good map and 4-inch (100-mm) telescope. Omega Centauri is the brightest globular cluster (see page 124) in the sky and looks like a small fuzzball with the naked eye. Give it a try! Source: Stellarium*

Discovered by the European Southern Observatory in 2016, Proxima Centauri b is about 1.3 times more massive than the Earth and revolves around the star once every 11 days at a distance of about 4.6 million miles (7.5 million km). Although it lies within Proxima's habitable zone, a distance at which liquid water would be stable on its surface, occasional but deadly blasts of ultraviolet radiation from flares on its host star may make this nearest exoplanet uninhabitable.

Facing south from Alice Springs, Australia, 10 p.m. June 1

▲   *From Down Under, Alpha Centauri, Crux and the southern Milky Way are seen at their best. The Magellanic Clouds (see page 150) are nearby satellite galaxies of the Milky Way. Source: Stellarium*

Imagine looking back toward the Sun and solar system from Proxima's planet. What would the sky look like at night? You might be surprised to learn that most constellations would appear virtually identical. Four light-years is a drop in the bucket when it comes to stellar distances, like visiting a neighbor down the block. How odd to think that after traveling so far, the sky would appear nearly unchanged. To see a more pronounced shift in the stars, we'd have to journey scores to hundreds of light-years from home.

Viewed from Alpha Centauri, the Sun would be the brightest star in the constellation Cassiopeia, the one shaped like the letter "W". If alien Centaurian eyes scan the skies, we're a notable light on their star maps.

How to find Alpha Centauri

There's a sure and easy way to find our featured star. If you're traveling to the tropics in search of the Southern Cross (see page 48), look about a fist-length to the left or east of the cross for two side-by-side bright stars three fingers apart. The one closer to the cross is Hadar or Beta Centauri, and the other, Alpha. These twin gems are sometimes called the **Southern Pointer Stars** because they point to Crux.

Alpha culminates two hours *after* the cross. To see the star at its highest, look due south in early April around 2 a.m. local daylight time, early May around midnight, early June around 10 p.m. and early July around 8 p.m. You'll need clear skies and the ability to see down to the southern horizon to nab it if you're observing from the southern extremes of the United States or the northern Caribbean. As always, the farther south you travel, the higher and brighter the southern stars appear above the horizon.

RESOURCES

- Latitude finder: latlong.net

- How to find the south celestial pole: oneminuteastronomer.com/3307/finding-south-celestial-pole/

- Online Planetarium—The Sky Live, an excellent, interactive star map of the entire sky from anywhere on Earth. Just select your city and you're set! theskylive.com/planetarium

- Stars are ranked in brightness according to magnitude. The higher the number, the fainter the star. Learn more here: stargazing.net/david/constel/magnitude.html

13

Orion Nebula

In kindergarten, I always looked forward to the afternoon nap. We'd lay out our blankets, and the teacher would draw the shades then turn off the lights. For the next half-hour, the two dozen of us snuggled into our little cocoons and slumbered in the semi-darkness. I sometimes think of that quiet time of regeneration when I eyeball the Orion Nebula on winter nights.

Tucked inside the puffy folds of this glowing cloud of interstellar dust and gas, hundreds of embryonic stars toss and turn as they prepare to shed the clouds that shroud them and burst forth as blazing new suns. A few of them are already up and about, spilling enough light for us to see the nebula as a small star-studded puff directly below Orion's Belt.

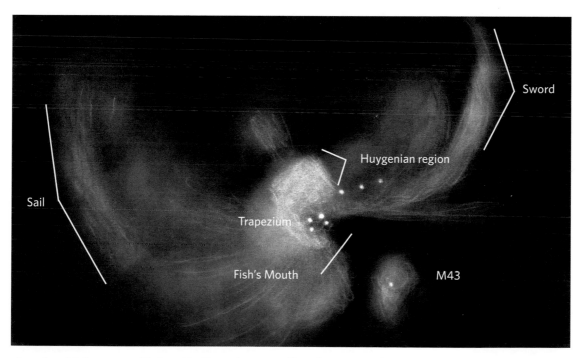

▲ *Small telescopes will show the Trapezium as well as the texture, shape and even a little of the nebula's green coloration. Larger ones, 8 inches (200 mm) and up, reveal more details and subtle red hues as shown in this sketch. Credit: Bob King*

Hundreds of nebulae dot the sky, but the Orion Nebula ranks among the very brightest. Easily visible with the naked eye from suburban and rural locations on Moon-free nights, it's a hazy patch about two full moons (1°) wide centered on the bright multiple star, Theta Orionis, also called the Trapezium. The nebula stretches across 25 light-years, the same distance from Earth to the bright star Vega of Summer Triangle fame, but at a distance of 1,345 light-years, fits cozily in the field of view of a small telescope.

Maybe the most amazing thing about this largest and closest star-making factory is the age of the stars that set it aglow: some are as young as 30,000 years! Early modern humans in Europe were at work on cave paintings before the nebula was even visible to the eye. Not long before people crossed the Bering Land Bridge that once connected Asia to the Americas, the first stars in the nebula burned and blew their way through billows of dust, exposing themselves to the gaze of humanity for the first time. We've grown up with the Orion Nebula. As it has evolved, so have we.

In the twenty-first century, the Orion Nebula remains a hotbed of research as astronomers seek to understand the process of stellar birth by the hand of gravity from simple beginnings as gas and dust. As skywatchers, we can partake of both the science and raw beauty of this most famous of nebulae. Through binoculars, Orion reveals additional stars and a shape that suggests a flower opening in morning sunlight. An 8-inch (203-mm) telescope brings this inanimate gas cloud to life and adds color, too. Green is easiest to see with hints of rose-red in the nebula's outer furls.

▲ *The Orion Nebula is an interstellar birthing center where stars by the hundreds incubate in dusty cocoons. Spanning about 24 light-years, it's about 1,350 light-years away. Source: NASA/ ESA/HST*

Similar to how neon gas glows orange-red when an electric current passes through it, the hot, young stars of the Trapezium, especially the brightest, Theta[1] Orionis C, beam powerful ultraviolet light across the nebular gases and excite them to glow.

Through our telescopes, we see only a handful of Orion's stars. Most are hidden within their embryonic clouds of gas and dust and invisible in ordinary light. Astronomers use telescopes sensitive to infrared light—a form of light we feel as heat—to peer through the dust and photograph the fury of stellar birth. They've discovered nearly 3,000 stars holed up inside, making Orion as much a star cluster as a nebula.

Facing south around 11 p.m. January 1

⋀   *The nebula is faintly visible with the naked eye and easily found just below Orion's Belt. Source: Stellarium*

Millions of years from now, when the remaining gas is converted into stars and cleared by stellar winds, humans will look up at a brand new star cluster in place of the misty mass—the Orion Star Cluster. From the looks of it, it could become one of the richest and brightest clusters in the sky. Would that we could live so long to see it.

How to see the Orion Nebula

Make your pilgrimage to the Orion Nebula in the late fall and winter months when Orion climbs high in the Southern sky. But if you like viewing in shirtsleeves, you can start as early as mid-September, when the constellation stands high in the southeast before dawn. Through fall, it rises earlier and earlier, working its way into the evening sky, so that by November, the nebula's well-placed by 10 p.m. and due south and highest at 2 a.m. Consult the timetable to pick a time best suited to your schedule. The times listed are when Orion culminates or reaches its highest point in the sky:

- December 1—around 1 a.m.
- January 1—around 11 p.m.
- February 1—around 9 p.m.
- March 1—late dusk

Under suburban and rural skies, the nebula is visible with the naked eye as a "smudgy star" about 5° (three fingers held together at arm's length) directly below the center star in Orion's Belt. Binoculars will show it very clearly as a small cloud dotted with several bright stars.

For the most memorable view, I recommend a telescope. A modest 6-inch (152-mm) instrument will reveal all the basic features including color, shape and the Trapezium, but the sight through a 10-inch (254-mm) telescope or larger scope under country skies will knock you over. Guaranteed. Because that's more of an investment, check to see if there's a local astronomy club that offers public viewing sessions. That way you can view this cosmic wonder in a variety of different telescopes. But if you decide to purchase a larger but more affordable instrument for this and other wonders in our must-see list, see the suggestions and vendors on page 216.

RESOURCES
- How to visualize the Orion Nebula in 3D: skyandtelescope.com/observing/see-orion-nebula-3d12172014/
- Trapezium Cluster: en.wikipedia.org/wiki/Trapezium_Cluster

14

Carina Nebula

If the Orion Nebula had a big sister, her name would be Carina. Sometimes, it seems southern latitude skywatchers have all the good stuff. Not only do they get fantastic views of Orion, but they have the closest star (Alpha Centauri), the brightest and biggest globular cluster (Omega Centauri—a separate entry) and the largest and brightest nebula—Carina. All the more reason to plan that trip to the tropics.

Visible without optical aid as a brighter patch in the southern Milky Way, the Carina Nebula is a stellar nursery like Orion but on a far grander scale. Using NASA's Chandra X-ray Observatory,

Facing south from San José, Costa Rica (10° N.), at 8 p.m. on May 1

▲ *The Carina Nebula sits smack in the middle of the Southern Milky Way a fist-length to the right (west) of the Southern Cross. Source: Stellarium*

astronomers have uncovered more than 14,000 stars in the Carina Nebula region. With a diameter of more than 200 light-years, it's eight times as large as the Orion Nebula and covers twice as much sky, about 2°. Located a little more than a fist-length to the west or right of the Southern Cross, it's easy to spot from a dark sky.

Even the view in binoculars will make you "ooh" and "ah", as the nebula resides in one of the most stunning parts of the Milky Way. The whole region glitters with stars and star clusters slashed by jagged lanes of dark, interstellar dust. In a 10x50 glass, "rivers" of dark stardust part the glowing nebula into several individual islands, the whole overlain by a blizzard of stars. For a compelling experience, point a 6-inch (152-mm) or larger telescope here. Let's hear from nineteenth century English astronomer John Herschel again, from *Results of Astronomical Observations Made During the Years 1834, 5, 6, 7 & 8 at the Cape of Good Hope*:

"It is not easy for language to convey a full impression of the beauty and sublimity of the spectacle which this nebula offers . . . ushered in as it is by so glorious and innumerable a procession of stars, to which it forms a sort of climax, and in a part of the heavens otherwise full of interest."

I still remember my first look at Carina in a 4-inch (101-mm) scope. Because of its great size, I had to use my lowest magnification to squeeze everything into one field of view. Stars glimmered everywhere, and I recall seeing several star clusters within and just beyond the bounds of the nebula. A wide, dark channel of cosmic dust cut across the middle then angled sharply to the

▲   *Gem of the southern sky, the Carina Nebula is four times as large and brighter than the Orion Nebula. Credit: Harel Boren/CC BY-SA 4.0*

northeast. Inside the brightest "island," the brightest star within the nebula, Eta Carinae, glowed orange. Just to its west, I could faintly make out the Keyhole Nebula, a small, dark cloud in the shape of a—you guessed it—keyhole.

Eta Carinae shines at fourth magnitude and is dimly visible with the naked eye, but as with so many astronomical objects, an ordinary appearance often masks extraordinary character. While appearing single in most telescopes, Eta is comprised of at least two extremely massive, hot stars, one originally some 200 times more massive than the Sun, and the other 30 to 80 times as hefty. If you're looking for a star or stars most likely to "go supernova", I'd put my cards on Eta Carinae. Some time in the relatively near future, the stars will run out of nuclear fuel. Without the heat and pressure of nuclear burning to hold back the crushing grip of gravity, the star will implode and then rebound outward, tearing itself to bits in a supernova explosion that astronomers predict will shine as brightly as Venus.

Not that Eta's been napping peacefully in recent years. In April of 1840, during an outburst that began in 1827, it briefly reached magnitude –1.0—brighter than any other star except Sirius. Then it gradually faded back, fluctuated a bit and has been brightening steadily again since the mid-twentieth century.

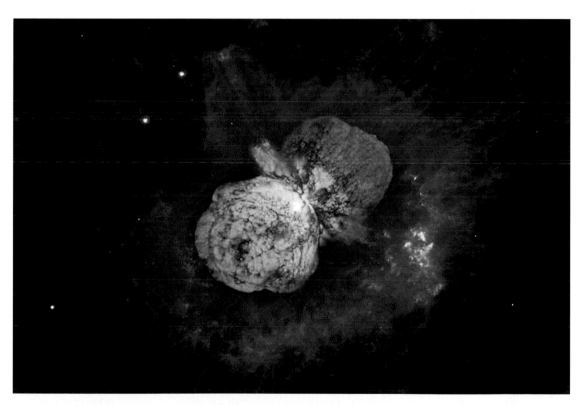

▲   *Nestled within the giant nebula is the much smaller Homunculus (Little Man) Nebula, home to the violently variable star, Eta Carinae. Astronomers believe the nebula formed during and after a major outburst of the star in 1840. Source: NASA/ESA/IST*

Back in the mid-1800s, Eta's extra brilliance changed how and what parts of the nebula were illuminated, so it looked different to observers then than it does now. Astronomers think all the tumult started when one or both of Eta's stars switched over to burning a different nuclear fuel. As steady and reliable as they seem to the casual gaze, stars evolve. They age just like people only over much longer time spans. Stars bloat up, shrink, toss off their atmospheres, heat up, cool off and even explode, all because of what fuel they have—or don't have—available to burn.

A medium-size telescope working at a high magnification of around 200x will show a tiny bubble of gas and dust called the Homunculus (Little Man) around Eta, released during the Great Outburst in the nineteenth century and still expanding outward at almost 1.5 million mph (650 kmps).

Whether Orion or Carina, pink clouds of ongoing star formation show up not only in the Milky Way but by the thousands in other galaxies, too. Within each, new stars are being fashioned from the same dust that built the Sun and solar system so long ago. You and I? We're part of a grand scheme that stretches from the Big Bang to the end of the universe as we know it, when some astronomers predict there will be little left but a thin gruel of particles and radiation. Such grim ends make one shudder but remind us how fortunate we are to be alive and have the opportunity to do good. Standing at the eyepiece of a telescope before the mighty Carina offers the opportunity to dip our toes in the river of time.

How to see the Carina Nebula

The nebula is located in Carina the Keel about 12° or a little more than a fist-length to the west of the Southern Cross. Spring is the best time for viewing, but because it's a misty blotch rather than a bright constellation, Eta Carinae is too low and faint from Key West. Hawaii offers an adequate view, but my recommendation is to travel farther south. From 10° north latitude, where Carina stands at least 20° (two fists high) and points south, will do it justice. Central and South America, Australia and Indonesia, among others, make ideal locations.

Best visibility times from 10° north latitude:

- February—from 11 p.m. to 2 a.m.
- March—from 9 p.m. to midnight
- April—from 8 to 11 p.m.
- May—from 8 to 10 p.m.

RESOURCES
- *The Little Book of Stars* by James Kaler
- Carina Nebula—Astronomy Picture of the Day: apod.nasa.gov/apod/ap131015.html
- Hubble's Carina Nebula: hubble25th.org/images/10

15

Pleiades in Binoculars

For some things, the naked eye is the best instrument—a meteor shower or a conjunction come to mind. For others, a telescope is essential. Think of Saturn's rings or the polar caps on Mars. Binoculars fit right in the middle and are perfect when it comes to bright star clusters like the Pleiades, better known as the Seven Sisters.

If you're not familiar with the Pleiades (PLEE-uh-deez), they're a little cluster of stars shaped like a miniature Big Dipper that climbs into the eastern sky on fall evenings. Most people can pick out the six brightest stars and maybe a seventh, but a closer look using averted vision will coax up to a dozen from a dark, moonless sky. The cluster looks misty as if seen through smoke or haze because fainter stars near or below naked eye visibility blend together to "fill in" the spaces between the brighter members. Or is that just my frosty breath getting in the way?

▲ *The Pleiades make their first appearance in the evening sky in early fall, when they rise ahead of the bright star Aldebaran and the V-shaped star cluster called the Hyades, seen here just above the treetops. Credit: Bob King*

The Seven Sisters, named for the seven daughters of the Greek gods Atlas and Pleione, is a true, gravitationally bound cluster of stars born together in a cloud of gas and dust 100 million years ago. They move together through space, orbiting the center of the galaxy 444 light-years from Earth. Counting Pleiades makes for an excellent test of both how dark your sky is and the keenness of your vision.

I encourage you to take this vision challenge when the cluster stands high in the sky on a moonless night. After you've finished, how about a little dessert? Treat yourself to one of the most wonderful sights in the sky by pointing a pair of binoculars at the cluster. While any pair will reveal additional stars, the extra light grasp of a 40 mm or larger glass will show about *ten times* the number visible to the naked eye and enhance their brilliance. Instead of a fuzzy mass, the cluster's brightest stars burn with blue-white fire. Their palpable, sun-like radiance makes them look like what they really are—spheres of fiery gas streaming heat and light into space with a ferocity we can hardly imagine. The members that outline the mini-dipper shine 40 to 1,000 times brighter than the sun.

One of the first things you'll notice through binoculars is an attractive arc of some eight stars that uncoils to the southeast of the cluster's brightest sun, Alcyone. It's my favorite part of the cluster, and because it resembles a centipede, that's what I call it. All unofficial of course. Only

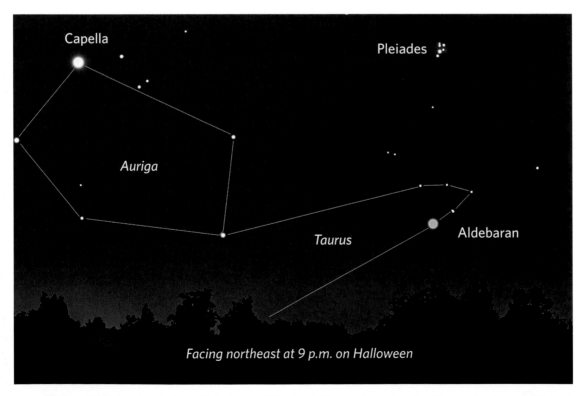

Facing northeast at 9 p.m. on Halloween

▲ *Trick-or-treaters can spot the Pleiades and Hyades in the eastern sky after they arrive home with their sweet loot. Source: Stellarium*

the International Astronomical Union (IAU) can officially sanction and approve star names, but asterisms like Orion's Belt and the Centipede belong to the common folk. So yes, feel free to dole out pet names to patterns that strike your fancy. Naming helps us characterize what we see and personalize our sky experience. Some names get passed around enough to stick and show up in guides and astronomy books. Examples, along with their catalog numbers, include the Eskimo Nebula (NGC 2392), the Fish Mouth (inside the Orion Nebula) and the Atoms for Peace Galaxy (NGC 7252).

There's something else I'd like you to look for using binoculars. Faint nebulosity enshrouds much of the star cluster. Starlight reflects off the tiny dust particles in the nebula the same way it does off cigarette smoke, lighting up the cloud. Before careful measurements proved otherwise, astronomers thought that the nebula was the remains of the Pleiades's birth cloud, but a study of the cluster's motion revealed that the two are unrelated. Like a traveling salesman, the Pleiades cluster is just passing through and will eventually leave its furry wraps behind.

The nebulosity appears brightest around the brightest stars, but it's faint overall except for a patch that hangs below the star Merope in the lower left corner of the Dipper. Resembling breath on a mirror, it's visible in 50 mm or larger binoculars under dark skies.

Sometimes on cold nights, I like to look up at the blue, luminous sister stars and imagine their heat reaching across the gulf of space to warm my hands and cheeks. Of course, it's all in my head, but the anticipation of the Pleiades return always warms the heart.

▲ The stars in the little-dipper-shaped Pleiades star cluster are named for Atlas and Pleione and their seven daughters, hence its more familiar name, the Seven Sisters. Credit: Bob King

How to see the Seven Sisters cluster

Bright enough to show from everywhere but badly light-polluted urban areas, the cluster makes its first appearance in late June at the start of dawn. We'll stick to easy evening viewing with times below shown for midmonth:

- October—eastern sky, 10 p.m. to 3 a.m. Culminates high in the southern sky around 3 a.m.
- November—eastern sky, 7 p.m. to midnight. Culminates at midnight.
- December—eastern sky, 6 to 11 p.m. Culminates at 11 p.m.
- January—eastern sky, 6 to 8 p.m. Culminates at 8 p.m.
- February—western sky, 7 to 11 p.m. Culminates at dusk.
- March—western sky, 8 to 10 p.m. Culminates before sunset.
- April—western sky, 9 to 10 p.m. Culminates before sunset.

RESOURCES

- Seven Sisters myths and legends: thesevensistersseries.com/myths-legends/4587870397
- How Many Pleiades Can You See? skyandtelescope.com/astronomy-news/many-pleiades-can-see10222014/

16

Moon Occultation

If you've never watched the Moon steal a star from the sky, prepare to be amazed.

The Moon moves about a fist-length to the east each night as it travels around the sky on or near an imaginary circle called the **ecliptic**, the path also taken by the Sun and planets. It completes the circle in just under a month and then begins the journey again. The ecliptic slices through the twelve constellations of the zodiac with names familiar to us from astrology and birthdays such as Aquarius, Scorpius, Pisces and Gemini. Four of the twelve possess a bright star: Scorpius has Antares; Taurus, Aldebaran; Leo, Regulus; and Virgo, Spica.

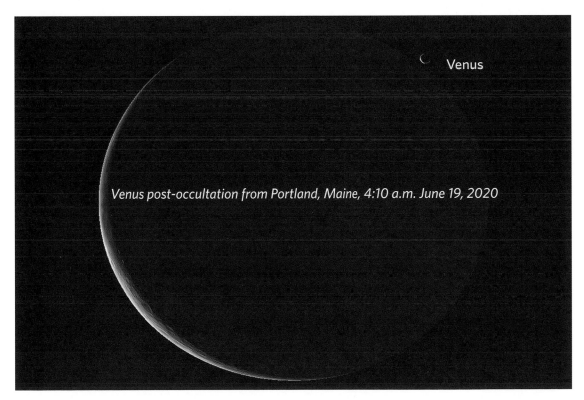

Venus

Venus post-occultation from Portland, Maine, 4:10 a.m. June 19, 2020.

▲ *Observers in the far northeastern U.S. will see side-by-side crescents when Venus returns to view after its occultation by the waning crescent moon on June 19, 2020. Source: Stellarium*

All four stars lie close enough to the Moon's path that they're occasionally covered by our satellite in an event called an **occultation**. Because they're bright, occultations of the four are sometimes viewable with the naked eye, especially if the Moon's a crescent and its dark edge covers the star. Other times, binoculars or a small telescope are necessary to clearly see the Moon blank the star from view.

The Moon travels along its orbit at an average speed of 2,288 mph (3,683 kmh). That's four times faster than the cruising speed of a 747. But because it's 239,000 miles (384,000 km) away, it appears to slowly creep eastward from night to night. As it approaches a bright star, say Spica, the two get closer and closer until Spica appears to hover at the Moon's edge for what feels like

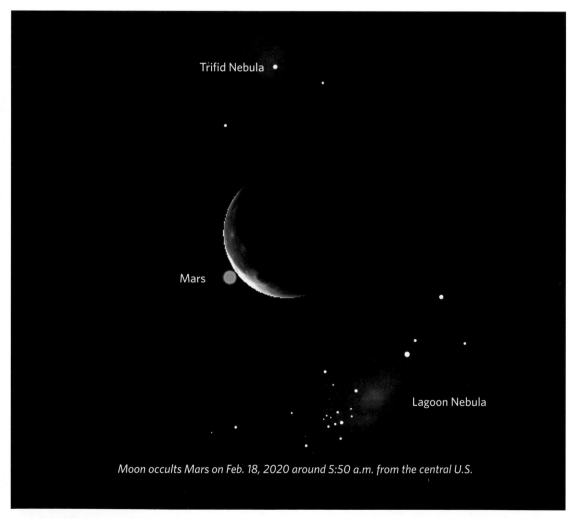

Moon occults Mars on Feb. 18, 2020 around 5:50 a.m. from the central U.S.

⌃   *The waning crescent moon occults Mars for much of the U.S. and parts of Canada and Mexico at dawn on Feb. 18, 2020. At the time, the Moon sits squarely between two bright nebulae in Sagittarius, the Trifid and Lagoon. Source: Stellarium*

forever. Will it ever disappear? And then all at once, Spica's gone in a flash! Minutes up to more than an hour later, the Moon slides away from the star, and it snaps back into view along its opposite edge.

The suddenness and brevity of an occultation will take your breath away. You'll also learn something important about the Moon—it has virtually no atmosphere. Of course, you already knew this. But to see it live before your eyes is to experience the real thing. If it had air like the Earth, the star would gradually fade as it got closer and closer to the Moon's edge. That's not what happens. Instead the star blinks out in an instant.

By observing an occultation, we also get a sense of how far away stars are. Even a star dozens of times bigger than the Sun blinks away in an instant, as if it were a mere pinpoint, which it practically is given its tremendous distance. Finally, you can directly experience the Moon's motion through space as the star makes a fleet retreat behind the lunar limb or pops out the other side.

Astronomers use occultations to uncover stars that appear single but have an extremely close but otherwise invisible companion star. The Moon's edge acts as a finely-honed knife blade that "slices" these hidden doubles into two. When a star scrapes along the very edge of the lunar disk, flashing in and out of sight in a **grazing occultation**, you're really in for a treat. Back in the day before orbiting lunar spacecraft, amateur and professional astronomers used these rare events to map the profile of the Moon's limb with its pointy peaks and crater walls alternating with low-lying valleys.

Occultations occur at particular times for particular regions, so an event in Europe might not be visible in the United States. Or the Moon may occult a star from the southern United States and glide just beneath it for skywatchers in the northern states. Although lots of stars will be occulted in the coming years, there will be no occultations of the four bright stars described earlier until the evening of August 24, 2023, when the first quarter moon will cover Antares in Scorpius, the first in a series that ends in 2028. Spica occultations follow in mid 2024, Regulus in 2025 and Aldebaran in 2033. Peering into my crystal ball, I've found several of the best visible from North America and Europe/Asia in the upcoming decade.

Stars aren't the only objects to cross paths with the Moon. Planets make great occultation fodder, too. They're fascinating in their own right because planets have disks (they're not pinpoints), so it takes the Moon seconds to almost a minute to creep over and devour one depending on how large it appears. It's also wonderful to see such a close juxtaposition of the Moon and a distant object like Saturn or Jupiter and know that for a time, it's been eclipsed from view and gone from the night sky.

How to watch an occultation

Use the table and simulations below to know when and where to view the events. All of the occultations listed are visible with binoculars, but consider training a telescope on them for a steadier view. Be sure to go outside about 15 minutes before the start, so you can get oriented, then just look through your telescope and watch the Moon slowly creep up to the star or planet. I've only listed the brightest events; check the resource links for many more and for specific times.

Occultation highlights

- February 18, 2020, start of dawn in the eastern sky. The bright limb of the waning crescent Moon occults Mars.
- June 19, 2020, start of dawn very low in the northeastern sky. Venus emerges into view along the crescent Moon's dark limb shortly after rising. In the United States, only visible in the far northeast.
- August 25, 2023, evening in the southern sky. The first quarter moon occults Antares visible across North America except the far north.

RESOURCES

- Bright star occultations: lunar-occultations.com/iota/bstar/bstar.htm
- Diary of astronomical events: poyntsource.com/New/Diary.htm

17

Moon Dogs

Have you ever seen a ring around the Sun or Moon? Yes? Then you're one step closer to seeing your first **moon dog**. Rings or halos, as they're known, are fairly common once you get into the habit of looking for them. They form in **high cirrus** or **cirrostratus clouds**—the thin, feathery ones—when sunlight or moonlight is bent by billions of tiny, pencil-shaped ice crystals.

A halo is only one piece of a full ice crystal display, which, depending on the type and purity of the crystals and how they refract light, can be spectacularly complex with multiple halos; **parhelia**, better known as sun dogs; and all manner of arcs. When I was 14, I saw a multiple halo–sun dog display one March afternoon that moved me to fill an entire page of my sky diary with a pencil sketch of the sight, all carefully annotated.

Many of these same phenomena happen with the Moon, especially around the time of a full moon, when our satellite shines brightest. Even then, not all of these crystal creations materialize because the Moon is nowhere near as brilliant as the Sun. To make all this stuff, nature needs light, the more the better. That's why we'll limit this must-see to a single, moderately rare event— sighting moon dogs.

But don't let that stop you from trying to see everything ice crystals have to offer. Routine observation of the sky day or night will ultimately reveal everything to your eyes that you've read about in a book or seen online.

Light moves in a straight line at 186,000 miles a second (299,338 kmps) unless it's reflected or deflected. When a ray of sunlight passes through one of the six-sided, pencil-shaped crystals in a cirrus-type cloud, it's deflected not once but twice: first when it enters the crystal face and a second time when it exits. Most of the light that enters one of these microscopic crystals departs it at an angle of 22°. Now, multiply that crystal by billions of others just like it and you get a bright, narrow circle of light with a 22° radius, the distance from the Sun or Moon to the halo's edge.

With so many ice crystals up there, why doesn't a halo look like a big bright blur around the sun? Why so perfect? It's true that all the crystals up there are deflecting light in the same way, but our eyes only receive sunlight from those crystals in the right positions to deflect sunlight *our* way, and those positions line up to make a circle.

So far, so good. Because a circle's radius is only half its diameter, a complete lunar or solar halo spans twice that or 44°—more than four fists of sky. That's *big* and the reason both solar and lunar halos look so impressive.

▲ *A lunar halo ornamented with colorful moon dogs makes for a magical winter night. Credit: Sebastian Saarloos*

Not only do bits of ice deflect or **refract** white light, they also cause it to fan out into its component colors, similar to what a prism does. Each color of light is refracted to a slightly different degree. Red light is refracted less than blue, tinting the halo's inner edge red and its outer edge blue. Sometimes, you can see hints of yellow and orange between the red and blue, but because these colors overlap, it can be difficult to separate them out.

Cirrus and cirrostratus clouds are also home to **plate crystals**. Imagine slicing up a pencil crystal like a nice salami into thin, flat plates. If the plates are floating randomly, they contribute to the ordinary halo, but if their flat faces are nearly horizontal to the ground, the double refraction of light through one face and out the other concentrates sunlight into two bright patches of light on either side of the halo level (or nearly so) with the Sun.

We call them sun dogs because they accompany the Sun like our favorite, furry companions would accompany us on a walk. Some even have tails. No kidding. Light exiting the plates at angles other than 22° creates a smoky, pale-blue tail extending outward and away from the sun dog. Most sun dogs are subtly colored for the same reason a halo displays color—each hue is refracted to only a slight degree, with the colors often overlapping to make white. Like a halo, you'll see red along a sun dog's inner edge and pale blue on the outside. Sometimes, a sun dog shines almost as brightly as the Sun and in vivid color. You might be driving to or from an errand and be stunned by the sight through your windshield. These rare displays usually last just a few minutes and have sometimes been confused with UFOs.

Moon dogs are identical to sun dogs in every way except they're infrequent. That's because you need lots of light to see a moon dog, and that only happens during the approximately week-long period centered on a full moon. You're not likely to see color in moon dogs either because moonlight's no match for sunlight. Our eyes need more light before the retina's color receptors kick in. Fortunately, cameras are up to the task. Experiment with time exposures of 2 to 20 seconds with your lens wide open (set to f/2.8, 4 or 4.5 depending on your equipment) and the ISO set to 400 or 800.

High, thin cirrostratus clouds often mark the start of a cloud sequence that signals the approach of a warm front and rain or snow. If you see a halo and moon dogs, there's a fair chance it may rain within a day. Before more advanced forms of weather forecasting, this bit of cultural wisdom became the proverb "when halo rings the Moon or Sun, rain's approaching on the run."

How to see moon dogs

Make a habit of looking up on bright Moon nights especially when rain or snow is in the forecast or when thin, filmy clouds soften the Moon's outline. Lunar halos are nearly as common as the solar variety; you'll probably see a dozen before spotting your first set of glowing hounds. Moon dogs resemble a pair of eyes looking down on Earth from their chilly heights. Seeing them reminds us how the smallest things combine forces to create sights that will stop you cold.

Speaking of which, moon dogs or sun dogs can form during very cold weather when ice crystals in the air slowly drift downward with their flat sides parallel to the ground. On these bitter days and nights, you can even see the crystals—called **diamond dust**—glinting in the light right before your bare eyes. Nature never wastes an opportunity to finger paint with ice.

RESOURCES

- Moon/Sun dog formation: atoptics.co.uk/halo/dogfm.htm
- National Weather Service forecast for where you live. Enter city or zip code: weatherservice. co/us/1/?keyword=national+weather+service+forecast

18

Distorted Moons

Beauty can be many things. A moonrise is one of them. Moonrises also excel at demonstrating how Earth's atmosphere interacts with and alters the appearance of celestial objects. Unless you're an astronaut, every time you look up at the stars, planets, Moon and Sun, your line of sight passes through miles and miles of air. In photos taken from space, the atmosphere looks like an insubstantial bluish mist, but it's critical to keeping us alive and the source of so many wonderful astronomical illusions.

Twinkling stars, the oval rising Moon, sunset reds and oranges, and the dimming of stars near the horizon are just a few ways the atmosphere affects how we see the sky. Even the simplest thing—a blue sky— is caused by tiny molecules of air that scatter away blue light from sunlight and send it bounding across the sky. With the blue removed, the Sun appears pale yellow to our eyes.

Air thins with altitude, so the higher up you look, the less air you look through. The bottom 10 miles of atmosphere, called the **troposphere**, contains 80 percent of the total atmosphere. That's where the air is densest. A box of air a meter on a side (39.4 inches) weighs 2.7 pounds (1.2 kg) at sea level but only 2 pounds (900 g) at 10,000 feet (3,050 m) up.

When you look at something near the horizon, your gaze cuts across hundreds of miles of the thickest, densest air at the bottom of the atmosphere. The closer the Sun (or Moon) is to the horizon,

▲   *The rising waning gibbous moon experiences fantastic distortions as its light passes through layers of air of different temperatures (and densities). Credit: Harald Wochner*

the more air its light must pass through and the more colors that are scattered away. By sunset, blues and greens are gone, with only yellows, oranges and reds able to blast past the molecules and make it to our eyes. We can't help but remark: "What a beautiful sunset!" It's also some of the "dirtiest" air, littered with natural materials like dust, sand, salt, volcanic ash, aromatics released from forests and fields and human-made pollutants such as soot, car exhaust and sulfur dioxide.

Except during the large volcanic eruptions, this material typically mutes the color of the rising or setting Sun and Moon and dilutes the hours of twilight. Dense air also acts like a prism and bends (or refracts) the light that passes through it similar to how water bends or refracts the light. It's called **refraction**.

Anyone spear hunting for fish has to take refraction into account. Light reflecting off the fish and up to your eyes gets bent or refracted when it exits the water, arriving at a different angle from whence it came. If you're not aware of this, you'll always aim the spear at where the fish *appears* to be and not where it actually is. Whenever light travels from one medium to another or from a denser part of a single medium to a less dense part, it gets bent.

Air near the horizon excels at refracting light, and you can guess why: It's denser than the air immediately above it. The iconic oval Moon on the rise is shaped by refraction. Denser air close to the horizon refracts or "lifts" the bottom of the Moon more than the top, "squeezing" it into an oval. The Moon looks less squished the higher and higher it climbs in the sky because the air along our line of sight thins rapidly and refraction effects lessen.

Ready for more weirdness? Refraction is so strong at the horizon that it lifts the entire Moon or Sun into view before either has actually risen. If we could magically remove the atmosphere at the moment the Sun's bottom limb breached the horizon, our star would suddenly disappear! If we then waited for the *real* sunrise, again with no atmosphere, the Sun would come up perfectly round and glaring white. No pleasing warm hues to greet the day. Yet another reason to appreciate what the atmosphere does for us.

When you look across so many miles of lower atmosphere at a low Moon or Sun, changes in air density, shifting winds and turbulence can imprint their own special distortions on their disks, making every rising and setting a unique experience. Turbulence can cause pieces of the Moon to peel off and float away (seen best in binoculars), while variations in the degree of refraction with height or a temperature inversion can dimple or even cut "steps" into a rising or setting Moon, making it look like a Mayan pyramid.

Usually, air cools with altitude, but in a temperature inversion, the air at ground or water level is colder than the air above, inviting all kinds of weird distortions including superior mirages.

Seeing these mind-bending phenomena doesn't take fancy equipment. Most are visible with the naked eye though a few will show best in binoculars. One of those is a color effect called **dispersion**.

Earlier, we learned that white light is comprised of all the colors of the rainbow. Refraction at the horizon not only lifts up the entire Moon wholesale, but on closer inspection through binoculars, reveals that each color is lifted to a different degree, spreading out the rising lunar disk the same way a prism spreads white light into a rainbow. Blue is bent more than red, so one edge of the Moon is fringed in blue and the other shimmers red.

Finally, clouds add their own touch of wonder to the rising and setting Moon. They take on the Moon's ruddy colors and form menacing dark streaks across its face. One of my favorites is what I call the "lunar bulge." Sometimes, the top and bottom of the Moon's disk are hidden by clouds with light streaming from a narrow band across the middle. The contrast between light and dark makes the bright part bulge out and distort the Moon's shape.

How to see warped and distorted moons

Go out regularly to watch full or waning gibbous moonrises and sunsets and sunrises, and don't forget to bring binoculars and a camera to record and share what you see. If the local TV weatherperson mentions a temperature inversion occurring for your area, that's a great time to look for odd atmospheric effects.

The hardest part might be simply finding a horizon. Many of us live where trees and buildings block the view. Large lakes, wide-open prairie country and mountaintops make great places to catch moonrises. Maybe you have one nearby? If so, take a drive there during the daytime and find a place you can return to at the next rising or setting opportunity.

RESOURCES

- Moon phases: timeanddate.com/moon/phases/

- Moonrise and moonset calculator: timeanddate.com/moon

- Sunrise and sunset calculator: timeanddate.com/sun

- Atmospheric optics: atoptics.co.uk/

- *Light and Color in the Outdoors* by M. G. J. Minnaert. The bible of atmospheric effects: amazon.com/Light-Color-Outdoors-Marcel-Minnaert/dp/0387979352

- *Color and Light in Nature* by David K. Lynch and William Livingston. Great photos and illustrations: amazon.com/Color-Light-Nature-David-Lynch/dp/0521775043

19

Light Pillars

If you don't mind the cold, you will become a good seeker of light pillars. Not that they always form in cold weather—all you really need are high clouds made of ice crystals to see the ones that tower over the Sun and Moon—but some of the most striking, even surreal pillars appear over artificial lights on bitter cold nights.

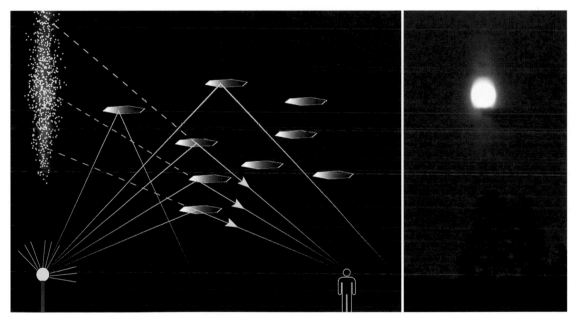

▲ *Light pillars form when horizontally-floating ice crystals act like tiny mirrors, reflecting light above and below a bright light source (here, an outdoor light) to form a vertical column of light. They can appear above both natural and artificial light sources. Source: Adis 1av/CC BY-SA 4.0*

A light pillar looks like a vertical band of light standing above (and sometimes below) the Sun or Moon. It might look like the Moon or Sun is shooting a beacon of light into the sky, but nothing could be further from the truth. A pillar isn't a *thing* but rather a collection of light glints reflecting off a particular set of ice crystals between you and the light source.

If you're out with a friend admiring a light pillar, each of you is seeing a slightly different version of the pillar because your lines of sight pass through a different batch of floating crystals. By the way, the same is true of all halo phenomena and rainbows. Each is unique to that person's viewpoint.

In a light pillar, billions of flat, hexagonal plate crystals in cirrostratus clouds drift horizontally (plate-side up) toward the ground. Light reflected from the bottom sides of all these tiny "mirrors" is reflected downward to the observer who sees a streak of diffuse light above the rising or setting Moon or Sun.

Not all light pillars look alike. Really tall spires form from ice crystals at higher altitude. The tallest completely change character, appearing as a cluster of short streaks of light radiating from the overhead point or zenith. Narrow, bright pillars form when the crystals are stable in the horizontal plane, neither tipping nor tilting. But in more turbulent weather, tilting crystals can make for more feathery columns. The most striking pillars are often seen at sunset/sunrise and moonset/moonrise when a warm front is approaching, bringing with it a veil of high clouds in the western sky.

Once the Moon climbs above the horizon, watch for a lower light pillar directly beneath it from light reflected *upward* from the top sides of the crystals. Sometimes, both upper and lower pillars are visible as a great, vertical column of soft light standing above and below the Moon in the eastern or western sky.

As gorgeous as natural light pillars are, poorly shielded artificial lights—the very ones that lessen our view of the night sky—can produce amazing displays of spiky and colorful pillars. Even car headlights do this. On extremely cold nights, plate crystals can form directly in the air as "diamond dust." Illuminated by yard and street lights, they create a phantasmagorical scene of multiple spikes of light pointing skyward like some ice-crystal aurora. If the temperature is below about 10°F (−12°C), and ice fog or light snow is forecast, make a point to go out and look for the phenomenon.

⋀   *On cold, clear nights, ice crystals can form and hover over light sources like these homes and headlights to create a spectacular display. Credit: John Ashley*

Driving home one night when the temperature stood at 10°F (–12°C), I noticed flecks of snow in my headlight beams. Once I got away from city lights, all the oncoming cars were decorated with a double set of tall and narrow light pillars created by their headlights. Because the pillars stood so high, I could easily tell when a car was approaching well before the headlights flashed into view.

One night, a friend called to alert me to a fantastic display of northern lights visible directly over the city of Duluth, Minnesota. I ran out for a look and saw an amazing sight alright . . . of light pillars. Diamond dust up to its old tricks again!

How to see light pillars

Listen to the weather and pay attention to the arrival of warm fronts. Their progress often starts with the appearance of high cirrus and cirrostratus clouds that soften the outline of the Moon or Sun. If the clouds make their way into your region around moonrise or moonset (or sunrise and sunset), keep an eye out for light pillars.

Winter is the best time for diamond dust light pillars not only above and below the Sun and Moon but above any bright, artificial light source. As much as I hate to say this, identify unshielded light sources in your city—malls, car dealerships, apartment buildings and parking lots make good sources. When the conditions are right, keep your eyes peeled.

Don't forget to bring a camera and tripod. If they look amazing to your eye, they'll look even better in a time exposure. Set your camera lens to its widest aperture (usually f/2.8 to f/4.5), ISO to 800 or 1600 and experiment with exposures ranging from 2 to 30 seconds.

RESOURCES

- Atmospheric optics: atoptics.co.uk/halo/lpil.htm
- Light pillar photos: bit.ly/2tVmnxM

Bright Meteor Shower

Every night I'm out under the stars, I hope to see a fireball. I suppose if I just sat still in a reclining chair and looked up, my odds would improve. As it is, I mix looking up with observing through a telescope, so my fireball life list isn't what it could be. Still, I rarely return indoors meteor-less.

Fireball is the term given to meteors as bright as Venus or brighter, the unforgettable ones that elicit those prolonged "w-h-o-a-s". According to the American Meteor Society, experienced observers can expect to see one fireball of Venus's brightness for every 20 hours of meteor observing on average. For fireballs of magnitude –6 (about 1.5 times the brightness of Venus), the time increases to 200 hours. Clearly, they're not common, which is why many of us align our meteor watching with annual meteor showers.

▲ *A fireball blazes below the Pleiades as a shard of comet or asteroid vaporizes in the atmosphere some 60 miles (96 km) high. Credit: NASA/Bill Dunford*

Meteor showers occur when Earth passes through the trail of dust left behind by a comet, or more rarely, an asteroid. Particles, called meteoroids, range between 0.04 and 0.4 inches (1 mm and 1 cm) across, and the average distance between them is 18 to 25 miles (30 to 40 km) in a dense shower.

A typical meteor-making meteoroid—the name given to space debris before it strikes Earth's atmosphere—is the size of a single bread crumb. And the distance between each nugget amounts to a 20-minute drive on the freeway. Despite nifty illustrations, meteor showers are thin soup. But speed makes all the difference. Earth is moving at more than 66,000 miles per hour (30 km/sec) around the Sun, and meteor streams roar along at similar speeds. When our planet tears through the debris, it completes that 20-mile (32-km) "drive" in about a second, scarfing up a healthy number of meteoroids in a surprisingly short period of time.

Speed matters in astronomy. If I took that chocolate chip and threw it at you, you'd hardly feel a thing. But take that same chip and speed it up to 71,000 mph (114,260 kph), the average encounter speed of a Leonid meteor shower meteoroid, and it would feel like a bullet. Speed increases an object's kinetic energy, also called the energy of motion, the reason why a small thing like a

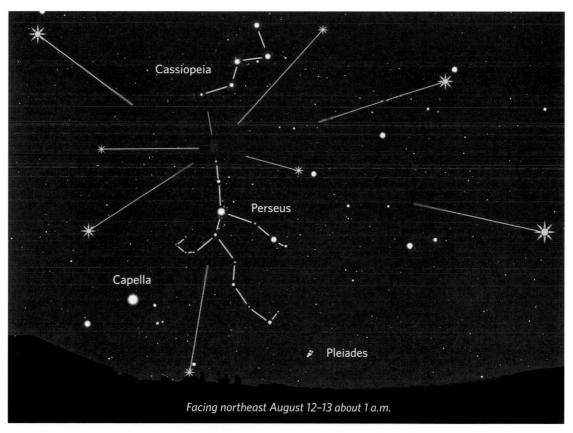

Facing northeast August 12–13 about 1 a.m.

▲ *The Perseid shower is the year's best known and easiest to watch because it happens during the summer, when many of us are outdoors at night. Source: Stellarium with additions by the author*

pebble-sized space rock or a lost screw can damage the hull or crack a window in the International Space Station as it orbits the Earth at more than 17,000 mph (27,300 kmh).

Because most material shed from a comet is tiny, the majority of shower meteors are faint, but there are plenty of middle-bright ones and almost always, a few jaw-droppers.

Dozens and dozens of meteor showers are known, but most are weak with counts of only a few per hour. Of these, nine are major meteor producers with rates ranging from 10 per hour for the Ursids up to 120 per hour for the Geminids and Quadrantids.

Shower names are based on the constellations from which the meteors appear to radiate, hence the Leonids (Leo), Geminids (Gemini) and so on. The Quadrantids name is unique, belonging to a now-defunct constellation between Ursa Major the Great Bear and Boötes the Herdsman called Quadrans Muralis the Mural Quadrant.

Of the major showers, the richest, most reliable and easiest to view are the Perseids, which radiate from Perseus just below the W of Cassiopeia in mid-August, and the Geminids, which peak on December 13 and 14 and stream from near the bright star Castor above Orion. The Perseids produce some 100 meteors an hour under dark, moonless skies at peak. The Geminids are

Facing east December 13 around 10 p.m.

⋀   *Geminid radiant—The Geminids, the strongest recurring meteor shower of the year, peaks on the night of December 13-14 averaging over 100 meteors from a moonless rural sky. Source: Stellarium with additions by the author*

currently the richest shower with a maximum of 120 meteors an hour under the same conditions. Remember that near a city, light pollution will cut those numbers in half. Add a bright Moon and the number halves again.

Even though the Quadrantids produce a rich display, timing has to be near perfect because peak shower activity lasts only a few hours. If those hours are from noon to 3 p.m. on January 4 for your location, you'll see very little. But if the peak occurs shortly before local dawn, when the radiant stands high in the northeastern sky, you'll get a great show. Hit and miss best describes this shower.

Meteor showers are best viewed when the radiant constellation is as high in the sky as possible, so most peak after midnight in the wee hours. Higher means fewer meteors cut off by the horizon. All showers send meteors across all directions of sky, but you can always identify a shower member by tracing its path backward. If it points back to Perseus, it's a Perseid. If not, it's a random or **sporadic** meteor.

Some meteor showers have outbursts that double or triple their expected output. This happens when Earth passes through a denser strand of comet dust laid down hundreds of years earlier that we're only encountering now in the twenty-first century. Meteor experts can often predict when these will occur, and you'll find them reported on websites such my own Astro Bob blog or on spaceweather.com.

The Leonid shower, which peaks every November 17 to 18 and averages 15 meteors an hour, puts on a much bigger show at 33-year intervals. That's how long it takes its parent comet, 55P/Tempel-Tuttle, to return to the Earth–Sun neighborhood. Solar heating vaporizes dust-rich ice from the comet. If Earth then crosses that fresh filament, a **meteor storm** results.

In 1833, a Leonid meteor storm rained meteors over the Americas at the rate of more than 100,000 per hour. Thirty-three years later, in 1866, Europe witnessed several thousand per hour.

Agnes Clerke, an American astronomer and writer remarked in *A Popular History of Astronomy During the Nineteenth Century*:

"On the night of November 12 to 13, 1833 a tempest of falling stars broke over the Earth . . . The sky was scored in every direction with shining tracks and illuminated with majestic fireballs. At Boston the frequency of meteors was estimated to be about half that of flakes of snow in an average snowstorm."

Michael Shiner, who chronicled events at the Washington Navy Yard, put it more succinctly:

"The Metors [sic] fell from the elements the 12 of November 1833 on Thursday in Washington. It frightened the people half to Death."

After a hiatus of a century, the showers returned with ferocity in 1966, briefly snowballing the atmosphere with 150,000 meteors an hour. The years 1999, 2001 and 2002 were all excellent Leonid years with 2001 the standout. That year my kids and wife and I spread a blanket out on the driveway at 2 a.m. and delighted to one fireball after another. Hands down it was the best shower I've ever seen.

Will the Leonids kick up another storm in 2033 or 2034? Unfortunately, as 55P/Tempel-Tuttle heads toward its next perihelion (closest approach to the Sun) in May 2031, it will pass near Jupiter in August 2029. The mighty planet's gravity will alter the comet's orbit in a way that will prevent Earth from getting close enough to sup on fresh material in both 2033/2034 and 2065.

The likelihood of storms improves in 2098 and gets really good in 2131, so at least the great grandkids will thrill to a meteor spree.

The rest of us will just have to take our chances with the reliable annual showers combined with as much "look up" time as we can manage.

How to see a bright meteor shower

Below are times when the Perseids and Geminids will be visible to best advantage under a moonless or near-moonless sky. Your job will be to find a place away from city lights to maximize your enjoyment of each shower. No special equipment is needed other than a comfortable reclining chair pointed east or south and a warm blanket or sleeping bag to huddle under to stay warm. Oh, and take your kids out or invite a friend. Meteors are best enjoyed when shared.

Perseids

Peak night August 12 to 13 but active two weeks before and after

- 2018: New moon. Ideal conditions!
- 2019: Waxing gibbous moon will compromise the shower until it sets about an hour before the start of dawn.
- 2020: Waning crescent moon in the morning sky will compromise the shower only a little. Still very good.
- 2021: Waxing crescent moon sets before things get underway. Ideal!
- 2022: Full moon will compromise the shower all night.

Geminids

Peak night December 13 to 14 but active several days before and after

- 2018: Waxing crescent moon sets early. Ideal!
- 2019: Full moon will compromise the shower all night.
- 2020: New moon makes for ideal conditions.
- 2021: Waxing gibbous moon compromises evening viewing but sets around 2 a.m. local time making the morning of the thirteenth ideal.
- 2022: Good evening viewing until the waning gibbous moon rises around 9 p.m. local time.

RESOURCES

- Meteor shower list: amsmeteors.org/meteor-showers/
- *Night Sky with the Naked Eye* by Bob King. More information about meteors and meteor showers.
- Fireball FAQ/American Meteor Society: amsmeteors.org/fireballs/faqf/

21

Jupiter's Galilean Moons

I'll never forget my first encounter with Jupiter in a small telescope. The bright "star" I'd been looking at for months became a creamy oval disk crossed by two dark stripes and accompanied by four tiny moons. The transformation, so easily accomplished with my little 2.4-inch (60-mm) instrument, got me hooked on Jupiter-watching for life. If you like space travel, you'll come naturally to a telescope—they're the poor man's rocketships to the stars.

Want to visit Jupiter, Mars, distant stars in the Milky Way? Plunk down your scope, aim and launch. Even city dwellers can make good use of a telescope when it comes to the bright planets, the Moon and Sun because they're mostly unaffected by light pollution. That's especially true with Jupiter, the brightest planet in the night sky after Venus. Despite being almost a half billion miles away, Jupiter's enormous girth and sunlight-reflecting clouds make it a standout among the eight planets.

Jupiter is so large—eleven times the diameter of Earth—if you could hollow it out like a Halloween pumpkin, every other planet in the solar system would squeeze inside with room to spare. That's why even a steadily held and sharply focused 10x pair of binoculars will reveal it has a shape, unlike the pinpoint stars.

Perpetual cloud cover also adds to the planet's shiny personality. Clouds reflect lots of light compared to the ground or water. Jupiter reflects about 50 percent of the sunlight it receives compared to Earth's 30 percent. While our planet has a healthy percentage of clouds, as we're all too aware on overcast nights, it's mostly covered in water with a modest contribution from land.

▲ *Through most telescopes, the four Galilean moons appear as tiny disks at best, but through the eyes of space probes, we can appreciate their unique characteristics. Source: NASA/JPL-Caltech*

The solar system's largest planet also has the largest family of moons, a total of 69 that we know about, as of mid-2017. Undoubtedly, more will be found in the coming years. Of that small army, only the four largest are bright enough to see in small scopes and binoculars. They're named Io, Europa, Ganymede and Callisto, and if you learn to say them in that order, you'll always know that's the same order of their distance from the planet, closest to farthest.

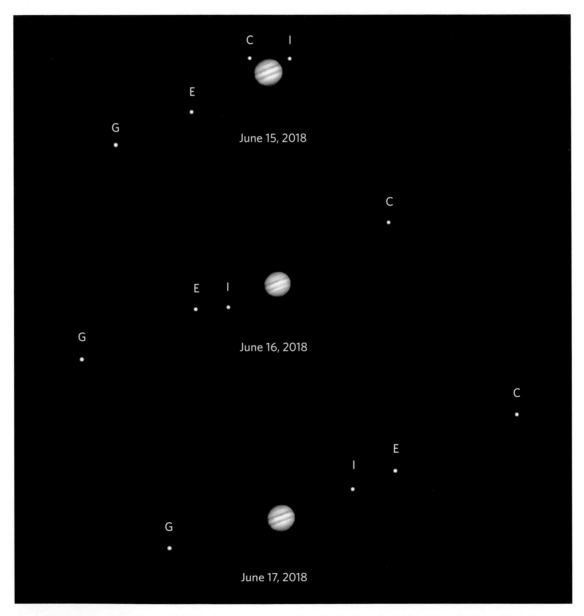

▲   *You can watch the dance of the Jovian satellites anytime Jupiter's in view. This example shows their changing positions over three nights in June 2018. I = Io; E = Europa; G = Ganymede and C = Callisto. Source: Stellarium*

Io's average distance is 262,000 miles (422,000 km)—a little more than the Moon's distance from Earth. Callisto orbits 4.6 times farther out or 1.2 million miles (1.9 million km) from Jupiter. All four are similar in size to our own Moon (2,160 miles/3,476 km). Europa's the smallest with a diameter of 1,945 miles (3,130 km) and Ganymede the largest at 3,273 miles (5,268 km).

Just as the Earth completes an orbit around the Sun in a single year compared to the twelve years it takes Jupiter, the closer moons circle the giant planet faster than the farther ones. When Galileo first saw Jupiter and family through his homemade, spyglass-style telescope, he was immediately struck by the scene's resemblance to a solar system in miniature and took it as proof that the Sun must sit at the center of ours. His revelation was in conflict with the prevailing view of an Earth-centered universe at the time, which got him into trouble with the Catholic Church. But that's another story.

As far as I know, no one gets in trouble for making remarks about Jupiter's moons nowadays. Instead, we enjoy visiting the planet any clear night it's out to revel at their diverse arrangements and other hijinks. One night, you might see all four strung out like socks on a laundry line, another night just two are visible with one hidden behind the planet and another in front but camouflaged by Jupiter's clouds. Sometimes, they gather in little groups, other times, they're spread far apart like wide-open arms. In the hundreds of times I've checked in on them, I don't think I've ever seen a repeat.

Each moon is unique. Io's surface is pepperoni-ed with some 400 volcanoes, about 150 of which are currently active. The lava they spew is rich in sulfur and coats the surface in crazy, pizza-like hues of orange and red. In a 6-inch (152-mm) or larger telescope, its color resembles a ripe peach compared to the blander hues of its brothers and sisters.

Europa, in contrast, is covered in a crust of water–ice at least 2 miles (3 km) thick. Beneath lies a salty ocean estimated at 62 miles (100 km) deep that's warm and mineral-rich enough to potentially support simple bacteria. That's why NASA will send its "Europa Clipper" to study the moon up close sometime in the 2020s. Of all the bodies in the solar system beyond Earth, Europa may be one of the most promising places to look for present-day life.

▲   *Jupiter and its four brightest moons are easy to see in almost any telescope. Even a pair of binoculars will show a few if you focus carefully and hold them steady. They were first seen by Italian astronomer Galileo in January 1610 and later named for him. Source: Stellarium*

Because of Jupiter's greater distance from the Sun, it's much colder there. So why are some of its moons still warm? Flexing!

As they orbit Jupiter, Io, Europa and Ganymede tug on each other, making their orbits slightly oblong, with one end closer to Jupiter and the other farther. On the closer end, Jupiter's powerful gravity pulls harder, deforming the moon. On the far end, the moon "relaxes" back to a more spherical shape. Repeated flexing creates a lot of internal friction, which heats up and partially melts their interiors. Water and magma result, some of which breaks through to the surface and either erupts as a volcano or freezes to form an icy crust.

Similar tugs from Earth flex the Moon's crust, while the Moon's gravity—often in concert with the Sun—flexes Earth's crust to the tune of 9.8 inches (25 cm) every high tide!

Ganymede is the brightest and largest of Jupiter's moons, and it may also have a salty, underground ocean. Because of its greater distance from Jupiter, only Callisto escapes the effects of flexing. Instead of geological activity and a relatively youthful surface, it's one of the oldest landscapes in the solar system, covered with more craters than any other place we know of. Even the Moon.

While we may see few of the details that make each moon unique, knowing a little about them will help illuminate that first night you swing a telescope toward Jupiter and discover these dancing satellites for yourself.

How to see the Galilean moons

First, you need to know where Jupiter is—an easy task if you're using a star-chart app on your phone. You can also use my opposition map to get a general idea where the planet will be the next few years, or head online to The SkyLive website. I also encourage you to download and fire up a copy of Stellarium for your mobile phone (also available for laptops and desktops).

With Stellarium, you can zoom in on Jupiter and see and identify which moons are out on a particular night. About every 7 years, Earth passes through the plane of the moons' orbits. As they circle Jupiter, we can watch them eclipse and occult one another in a series of **mutual phenomena**. This last happened in 2014 to 2015 and will next in 2021.

No matter when you watch, the moons occasionally pass in front of Jupiter and cast inky black shadows on its cloud tops in events called **shadow transits**. Other times, they tread into Jupiter's shadow and disappear right in front of your eyes in total eclipse. The moons also revolve around to the planet's backside and disappear or reappear at the planet's limb in events called **occultations**. To watch the fun, you'll only need a 4.5 to 6-inch (115 to 150-mm) telescope magnifying from 50x to 150x.

RESOURCES

- Stellarium or a phone app like Sky Chart described on page 10.
- The SkyLive/Jupiter: theskylive.com/jupiter-info
- *Sky & Telescope*'s Jupiter's moons (time of moon transits, eclipses and occultations): wwwcdn.skyandtelescope.com/wp-content/observing-tools/jupiter_moons/jupiter.html

22

Jupiter's Great Red Spot

Saturn has rings; Jupiter's got a big red spot. Each is iconic and yet transitory. It's very possible that Saturn's rings may have formed as recently as 100 million years ago, when a comet or small asteroid collided with one of its icy moons, the debris spreading and flattening into the pancake of ringlets we see today. Jupiter's Great Red Spot (GRS), nearly as famous as Saturn's rings, has been around since at least the mid-1600s though not unchanged.

You can think of the Great Red Spot as an off-the-charts hurricane some 30 times the size of a typical terrestrial storm with winds barreling up to 400 mph (644 kph). Sandwiched between jet streams blowing in opposite directions, the GRS spins counterclockwise once every 4 to 5 days.

Unlike earthly hurricanes, which are low-pressure storms, the Red Spot is a region of high pressure with clouds that stand high above their neighbors. One of the reasons this mammoth storm has lasted for so long is that Jupiter has no land for it to crash into and dissipate like a terrestrial storm. Once the GRS got rolling, it took on a life of its own.

If the planet had a chamber of commerce, its slogan might be, "We're all atmosphere, all the time."

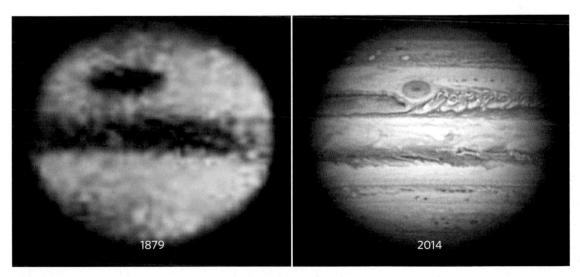

∧ Jupiter's all turbulent atmosphere, so the changing appearance of clouds is a given. In the late 19th century, the GRS was several times Earth's diameter and much easier to see in a small telescope. South is at the top. Credit: Left: Agnes Clerke's History of Astronomy in the 19th Century. Right: Damian Peach

The thickest, darkest cloud belts on Jupiter have a reddish hue but generally lack the intensity of the GRS. Exactly what causes the Spot's color remains a mystery. Jupiter's atmosphere is primarily made of the light gases, hydrogen and helium, with clouds of ammonia, ammonium hydrosulfide and water blown by jet streams into parallel cloud belts or bands.

Astronomers think that cosmic radiation and ultraviolet light from the Sun bombard and chemically alter ammonia and acetylene to create new compounds with a telltale red hue. Rising from below the cloud deck, they spread over the GRS like frosting on a cake. When exposed to sunlight or cosmic rays, the "frosting" turns red. The Spot's spin confines, concentrates and "cooks" the gases, maintaining the red color at least for a time. Long-time GRS watchers have seen its color change over the years from pale tan to peach to brick red. In my youth, the Spot's brick-red hue made it jump right out in a small telescope. Twenty years later, its color was closer to a pale tan and took more effort to see.

The variation in color is probably caused by varying amounts of acetylene and ammonia, their altitude and how long they've been exposed to sunlight.

The GRS changes in other ways, too. In the late 1800s, it spanned 25,000 miles (40,000 km), more than three times the diameter of Earth, and looked like a foot-long hot dog. Since then, it's contracted, become rounder and now measures only 1.2 Earths across. The shrinkage rate has been ticking up since about 2012, making for speculation that the GRS may disappear altogether in the not-too-distant future. Smaller eddies of swirling gases absorbed by the GRS appear to be sapping its energy and may be the cause of the downsizing.

▲ *NASA's Juno spacecraft photographed Jupiter's Great Red Spot during its seventh close flyby in July 2017. Although the Spot has been shrinking in recent decades, it's still larger than the diameter of Earth. Source: NASA/JPL-Caltech/SwRI/MSSS*

Will the GRS whirl away to nothing like so much water down the drain? Will our future skywatchers need to change its name to the Great Red Pimple? Always a possibility and all the more reason to get a look at this amazing feature while it's still around.

How to see the Great Red Spot

The best time to look is when Jupiter is conveniently placed in the evening sky during, and for several months after, the time of opposition. You'll find a list of oppositions below.

At opposition, Jupiter and Earth are closest for that year, and the planet and GRS appear largest. To see the GRS, a modest 6-inch (150-mm) telescope magnifying around 100x will do. Because Jupiter rotates once every 9.5 hours, you'll need to know when the GRS faces squarely in your direction. Technically, astronomers call it the time when the GRS transits Jupiter's central meridian. Transit times are published in monthly astronomy magazines such as *Sky & Telescope*, but they're also online at the websites listed below.

Jupiter's Red Spot Calculator is the easiest to use. Click on "Initialize to today" and then "Calculate," and a list of times (in Universal and your local time) pops up when the GRS is straight up in view. You can also type in a date of your choice and click Calculate. The GRS is well-placed for viewing an hour on either side of the time shown.

If atmospheric turbulence makes the planet blurry through your telescope, the GRS will be difficult to see, but if the air is calm, you should spy a pink eye staring back at you from Jupiter's southern hemisphere. For all its cyclonic fury in photos, it looks rather delicate in real life. Don't expect a photographic appearance. Do expect to marvel at a storm that's been around long before the United States became a nation.

If the image is crisp, increase the magnification to 150x or 200x for an even better view. You'll see other features, too, including the thick, dark North and South Equatorial cloud belts and maybe even the pale "bowl" in which the GRS resides called the Red Spot Hollow.

Jupiter oppositions and locations

- May 9, 2018 in Libra
- June 10, 2019 in Ophiuchus
- July 14, 2020 in Sagittarius
- August 20, 2021 in Capricornus
- September 26, 2022 in Pisces
- November 3, 2023 in Aries
- December 7, 2024 in Taurus
- January 10, 2026 in Gemini

RESOURCES

- Jupiter's Great Red Spot Calculator: shannonsideastronomyclub.com/jupiter_grs_calc1.htm
- Sky & Telescope Transit Times of Jupiter's Great Red Spot: skyandtelescope.com/observing/celestial-objects-to-watch/transit-times-of-jupiters-great-red-spot/

23

Mars's Polar Caps

Funny. Mars lacks Saturn's exotic appeal but has almost equal drawing power because we sense a familiar face. Eroded stream beds, buttes and mesas, sandstorms, starry nights and sunny days and, of course, those distinctive polar caps, help us feel a real connection to the Red Planet.

Has life evolved there like it did and continues to do on Earth? Or did it perish and leave only crumbles of bacterial microfossils behind? We may find the answer to these questions and others in the coming decade after more extensive study of the planet with probes that will scoop up and return promising rock and soil samples to the lab for study. Until then, we can look and wonder at Mars in our telescopes and try to imagine what it might be like to stand on its dry, rusty-dusty surface, arms extended in a receptive gesture as we take in the scene around us.

Two things stand out when observing Mars through a telescope—dark features on its surface called **albedo markings**, and if we're lucky, one or both polar ice caps.

Albedo refers to how much light a surface feature reflects back to your eye. Since much of Mars is covered in fine, iron-rich dust that reflects sunlight well, the planet has a bright pinkish-red appearance. Contrasting darker features are regions where the dust has been removed by

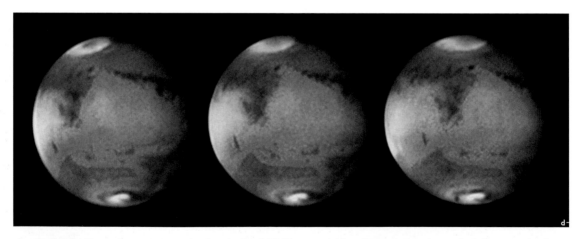

▲ *Both Martian polar caps were in view in April 2012, as was Mars's most prominent dark feature, the fang-like Syrtis Major. South is up in this sequence which shows Mars's rotation. Credit: Damian Peach*

prevailing winds to reveal the typically gray, basaltic rock beneath the planet's rusty overcoat. Basalt is a volcanic rock that's as common on Mars as it is right here at home. I pass millions of tons of ancient basalt every day on my way to work in Northern Minnesota.

The polar caps are deceptive. They mimic the appearance of Earth's frozen extremities, but they're composed of water–ice *and* carbon dioxide (dry) ice. Mars is plenty colder than the Earth because it's about half again as far from the Sun, the reason dry ice accumulates in the planet's polar regions. Earth's average temperature is 61°F (16°C) while that of Mars is a frigid –67°F (–55°C), more than 120° colder. Both caps contain permanent, year-round water–ice reserves and a seasonal cap of carbon dioxide ice that forms every fall and winter and then sublimates—goes directly from solid to gas—in spring.

Every fall, in each respective polar region, blustery clouds form and block the growing cap from view. Carbon dioxide ice freezes out of the air and covers the cap with dry-ice snow and ice, expanding its size. When spring arrives about 6 months later (Martian seasons last about twice as long as Earth's), the clouds part and the polar cap gleams brilliant white in the sunlight. Even a small 3-inch (76-mm) telescope will show the red desert planet capped with a white button.

As the season progresses, the cap shrinks as dry ice vaporizes directly into the thin atmosphere.

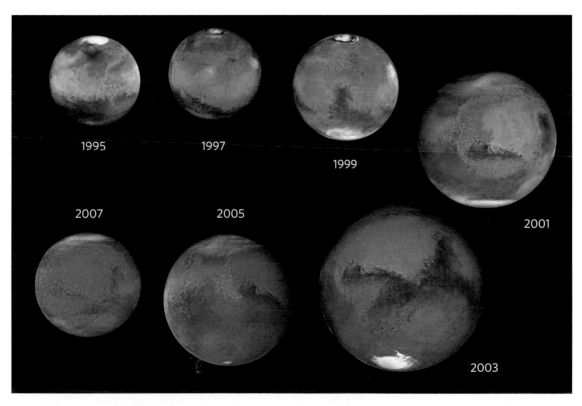

▲ *Color composite of Mars from seven of its previous oppositions, taken with the Hubble Space Telescope. Notice how first one polar cap and then the other is tipped toward the Earth. The planet's size also varies from one opposition to the next since its distance varies. Source: NASA/ESA/HST*

Meanwhile, the cap at the other end of the planet, where it's now fall, huddles under a massive blanket of clouds called the **polar hood**, which can be bright enough to make you think you're seeing the true cap. Occasionally, both caps are visible, but usually one is tipped toward the Earth and the other away during the times when the two planets are closest.

Mars is a small planet, only about twice as big as the Moon and half the size of Earth. To see features on its surface at their best, skywatchers concentrate their Mars viewing around the time of opposition when both Earth and Mars are lined up on the same side of the sun and closest together.

At their nearest, the two planets can be just 34.6 million miles (55.7 million km) apart; when farthest, they're separated by an odometer-busting 249 million miles (401 million km).

Mars's permanent north polar cap—the one that remains throughout the summer—is about 620 miles (998 km) across, while the southern cap is much smaller, only about 249 miles (400 km). You'll see one or the other best, when the planet appears largest and brightest.

For happy Mars viewing, I recommend a modest 6-inch (152-mm) or larger instrument. Throughout high school and college, I used a 6-inch (152-mm) reflecting telescope purchased with paper route money from my grade-school days. When Mars was close, I'd crank up the magnification to 250x or even 350x to eke out every albedo marking I could and watch the waxing and waning of the polar caps. I had the names of every feature on my fingertips and could rattle them off like a tour guide. Such are the joys of amateur astronomy, that intense observation of a planet or other sky object can practically turn you into a resident.

How to see the Martian polar caps

I've listed several upcoming Mars oppositions on page 93. Because Mars has a much more elliptical or "squashed circle" orbit compared to Earth's, opposition distances vary more than most planets. Some are closer, some farther. The 2018 opposition will be the closest since 2003 with Mars shining even brighter than Jupiter. The next, in 2020, will be almost as close and exciting.

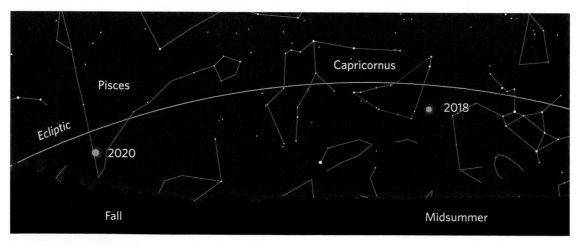

▲ *Mars appears largest and brightest at opposition. This map shows the planet's location around the opposition dates for 2018 and 2020. Source: Stellarium*

Oppositions

- July 27, 2018: Diameter: 24.3 arc seconds (abbreviated 24.3″) with the south polar cap tipped toward Earth. One arc second equals 1/60 of an arc minute; 30 arc minutes equals the apparent diameter of the full moon. Although 24.3″ sounds tiny, it's huge for Mars, a very small planet.
- October 13, 2020—22.6″—South Pole in view
- December 8, 2022—17.2″—South Pole and edge of North Pole
- January 15, 2025—14.6″— North Pole
- February 19, 2027—13.8″—North Pole
- March 25, 2029—14.5″—North Pole
- May 4, 2031—16.9″—North Pole but South Pole later in year
- June 28, 2033—22.1″—South Pole
- September 15, 2035—24.6″—South Pole

Telescope

I'd recommend a 6-inch (152-mm) or similar telescope equipped with an eyepiece(s) that allows magnifications of 150x or more; 150x is a nice trade-off between too high a magnification, which can make everything look blurry, and too little power to see these little white dollops. Eyepieces are sold by **focal length**, which is measured in millimeters. Low-power eyepieces have long focal lengths from about 25 to 40mm; high-power ones range from about 10 mm down to 2.5 mm. Look for an eyepiece with a focal length of around 5 to 8 mm.

If you can afford to spend the extra money, get one with a wide apparent field of view and good eye relief. You can easily find the magnification of a given eyepiece by dividing your telescope's focal length in millimeters by the focal length of the eyepiece. Look on the telescope tube or in the instructions to find its focal length.

Best viewing times

The poles are very tiny during the Martian summer and hidden by clouds in the fall and early winter, so late winter through early summer are the best times to see them. Below you'll find a list of times when the caps should be free of clouds and near their largest before their seasonal coats vaporize away. These times aren't set in stone! They're approximations based on past oppositions.

- May to July 2018—south cap
- April to July 2020—south cap
- February to June 2022—south cap
- January to March 2023—north cap
- November 2024 to March 2025—north cap
- October 2026 to March 2027—north cap
- November 2028 to February 2029—north cap
- January 2031—north cap
- August to October 2031—south cap
- June to September 2033—south cap
- April to August 2035—south cap

RESOURCES

- Stellarium or a phone app like Sky Chart described on page 10 so you know exactly where Mars is any night of the year.

- Mars oppositions through 2027: nakedeyeplanets.com/mars-oppositions.htm

- Mars calendar: planetary.org/explore/space-topics/mars/mars-calendar.html

- Mars opposition 2018: alpo-astronomy.org/jbeish/2018_MARS.htm

- Mars observers list. A dedicated group of Mars watchers with up-to-date information and photos to help you know what to anticipate when observing Mars: groups.yahoo.com/group/marsobservers

24

Eight Planets in One Night

First off, everyone gets a free planet. Yes, the very one you're standing on. Earth's always a great place to start. From here, we can look up and see all seven of our brother and sister planets—Mercury, Venus, Mars, Jupiter, Saturn, Uranus and Neptune. Usually, they're in this or that part of the sky, some visible during the evening and others before dawn. But rarely, they bunch up so that we might see them all together, arrayed across the sky like steppingstones across a river.

Each planet orbits around the Sun in a different length of time depending on its distance. Saturn's far away and slow, taking 29.5 years to complete a loop around the sky. If you see the ringed planet in Virgo one year, it will return to the same place 29.5 years later. Jupiter accomplishes its circuit around the Sun and sky in 12 years and Mars in 687 days.

Venus and Mercury orbit *between* the Sun and Earth, so they swing back and forth from one side of the Sun to the other, spending most of their time in morning and evening twilight. Getting the planets to gather in a confined space is like getting your kids to put down what they're doing and come to the table for dinner.

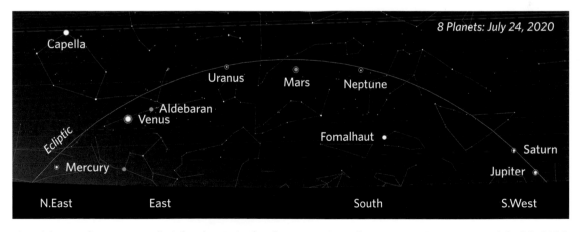

▲ *It's not often we see all eight planets in the sky at one time. One opportunity occurs on July 24, 2020. Be sure you're somewhere with a good view of both the eastern and western horizons. Clear skies! Source: Stellarium*

Of course, there's no "getting" planets to do anything. Instead, we simply have to wait for their diverse orbital periods to fortuitously align and bring them all into view at once. I've done the simulations and have listed below four times between 2018 and 2022 when all five bright naked-eye planets will be visible across the sky.

On three of the dates, Uranus and Neptune will also be out, making for a grand total of eight planets if you include the freebie. To see them all, you'll need reasonably dark skies, a pair of 40- to 50-mm binoculars (10x50s are perfect) and the provided maps. In all cases, the planet Mercury will lie near the eastern (dawn) or western (dusk) horizon, so find a place with a wide-open view *especially* in those directions.

The last easily visible grand alignment occurred in late January 2016, when skywatchers willing to brave the chill dawn could pick out *five* planets spread-eagle across 110° (eleven fists) of morning sky from west to east: Jupiter, Mars, Saturn, Venus and Mercury. The next grouping—a four-planet affair—occurs in late summer 2018 when Venus, Jupiter, Saturn and Mars fall along an 85° arc across the western evening sky.

The maps include an arc that runs near all the planets. This is the ecliptic, the plane of Earth's orbit as a line projected into space and traced out by the Sun once a year. The planets travel the same celestial highway, which traverses the twelve zodiac constellations, as they orbit the Sun. Why do they take the same road and never exit onto another? Because the planets essentially orbit in a flat plane like runners at a track meet. Earth's a runner, too. When we look across to see how the competition's doing, the other runners (planets) all appear to chug along at the same level and against the same backdrop. The level track is the ecliptic against the backdrop of the zodiac constellations.

Seeing all the planets spaced out across the sky affords us the unique opportunity to appreciate the essential flatness of the solar system. For a beginner stargazer trying to make sense of how the planets orbit, a multibody alignment beautifully illustrates the concept right before our eyes.

How to see five-planet and eight-planet alignments

Here are the four events and the times when we'll see them best. Sorry about the wait, but it's beyond my control. To explore these in more depth, install a star-charting app such as Stellarium and enter the dates of each. Then you can play around with times and visibility from your specific location. All of these events will be visible from midnorthern latitudes. In the first two, it will be more difficult to see Mercury if you live north of latitude 45°N because of an earlier start to morning twilight.

- Late July to mid-September 2018, dusk: Venus, Jupiter, Saturn and Mars form an arc from west to southeast that varies from 150° long (mid-July) to just 85° in early September.
- Late July 2020, best on July 24, dawn: Visible from about 1.5 to 1 hour before sunrise. Planets will be lined up along a 170° arc from east to west in this order: Mercury, Venus, Uranus, Mars, Neptune, Saturn and Jupiter. Jupiter will be very low, so seek a western horizon.

- Late April 2022, dawn—from east to west: Jupiter, Neptune, Venus, Mars and Saturn lined up in an arc about 32° long across the eastern sky. The crescent moon travels below the planets from April 24 to 27.
- June 22 to early July, 2022, dawn: The best! All seven planets and the Moon will appear in this order—Mercury, Venus, Uranus, Mars, the Moon, Jupiter, Neptune and Saturn—as they trace a 104° arc across the eastern and southern sky.
- Late December 2022, best on December 20, dusk—visible about an hour after sunset in a 148° arc from west to east: Venus, Mercury, Saturn, Neptune, Jupiter, Uranus and Mars. Because of bright twilight, Neptune won't show in binoculars until after Venus sets.

RESOURCES
- Stellarium or an app like Sky Chart described on page 10.

International Space Station

You haven't lived unless you've seen the International Space Station (ISS) fly over your home or apartment at 17,150 mph (27,600 kmh). OK, maybe I'm exaggerating a little, but not too much. Two hundred years ago, a giant spacecraft carrying men and women around the Earth every 90 minutes would have been unthinkable. Pure fantasy. And yet there it is: a bright torch reminding us of what's possible with a little imagination and a lot of sweat. With all the resources available including websites, email alerts and free mobile phone apps, it's never been easier to spot this amazing, manmade creation, the biggest satellite in the sky.

▲   *The ISS orbits about 250 miles (402 km) above the Earth's surface, circling the planet once every 90 minutes. It's the largest and brightest manmade object in orbit. Source: NASA*

As big as a U.S. pro football field including the end zones and weighing 462 tons (419 metric tons) or the equivalent of 231 midsize SUVs, the space station circles the Earth at an average altitude of 248 miles. That may sound like a long way up there, but that's considered low-Earth orbit or LEO. Satellites in high-Earth orbit, like those that relay communications around the globe or keep track of the weather, travel nearly 100 times higher.

The station was designed to support a crew of six but in a pinch, more souls can be accommodated. The fullest house as of this writing was thirteen, when the seven-member crew from space shuttle Endeavour docked with the six-person crew aboard the ISS in July 2009. No matter when you spot the space station, know that there's more than likely a handful of astronauts aboard. I still can't get over the fact that every hour and a half, our fellow humans are speeding overhead at more than 22 times the speed of sound.

As they hurry around the curve of the Earth, witnessing sixteen sunsets and sunrises a day, the astronauts run a host of science experiments in the unique "microgravity" environment of the station. They clean, check equipment, make repairs, prepare meals, conduct spacewalks to install equipment on the outside of the ISS or to make repairs. Crewmembers must also exercise two hours a day to stay fit and prevent bone and muscle loss that occur in a weightless environment.

It's not all work. In their free time, astronauts can hang out in the multiwindowed cupola, a perfect place to read a book or gaze at and take pictures of the Earth below. Its 31.5-inch (80-cm) circular overhead window is the largest ever flown in space. Arrayed around it are six additional

∧ *It's fun and instructive to follow the International Space Station across the sky. In this 30-second time exposure, the ISS cuts through the handle of the Big Dipper while traveling from west to east. Credit: Bob King*

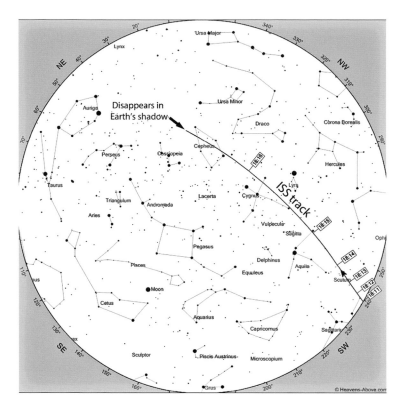

trapezoid-shaped windows. You'd have a hard time getting me out of there. Just one more photo?

Undoubtedly, you've seen pictures of astronauts floating around the space station, launching themselves from railings and fixtures using their hands like a fish might use its fins. Weightlessness looks like pure joy. The closest most of us come to that state is during sleep, when we occasionally dream of flying.

Weightlessness has nothing to do with a lack of gravity. The ISS remains in orbit thanks to the constant gravitational tug of the Earth below. Instead, it's all about falling. As it orbits, the space station is continuously falling toward the Earth. The reason it doesn't just burn up in the atmosphere after a few orbits is because it's also moving forward fast enough (remember—17,150 mph [27,600 kph]) to swing around the curve of the globe without hitting the ground. Seen from the ISS, the Earth ahead appears to curve away from the ship at the same rate that the ISS falls. Neither shall meet.

Freefalling causes weightlessness, but your mass—the stuff you carry around—remains the same. If you've ever been in an elevator that dropped too quickly to the next floor, you may have felt lighter on your feet for a moment—a taste of weightlessness. Essentially, the astronauts go about their work in a large, well-stocked, freefalling elevator.

Like many satellites, the ISS was launched from west to east to take advantage of Earth's rotation. At Cape Canaveral's latitude of 28.5°, Earth spins at 914 mph (1,471 kmh). That free boost means it takes less fuel to accelerate a satellite to orbital speed. Less fuel = lower cost. Closer to the equator, the Earth spins even faster, and the bonus is bigger. Farther north, Earth spins more slowly and completely zeroes out at the poles, where there's no free lunch.

▲　At the website Heavens-Above, you can view maps of any space station pass for your location. This example shows a pass over Philadelphia on November 28, 2017. The ISS is eclipsed by the shadow of the Earth and disappears from view before completing the pass. Source: Chris Peat/Heavens-Above

The launch direction directly affects where we look to see a pass. Without exception, the ISS first appears as a bright star somewhere in the western sky, whether that's southwest, due west or northwest. From there, it travels east across the sky until it either sets in the east or gets eclipsed by Earth's shadow first. A complete pass lasts about six minutes.

We see all satellites in a twilit or dark sky because they're high enough to still catch the rays of the Sun after the Sun has set for observers on the ground—like the proverbial mountaintop glowing sunset orange above the darkened valley below. But as time passes, the Earth's shadow creeps higher and higher above the eastern horizon as the Sun dips lower and lower below the western horizon.

During a pass in bright evening twilight, the ISS zips by unbothered by Earth's shadow. But 92 minutes later, when the shadow has climbed high enough in the sky to cover part of the station's path, something very interesting happens. The ISS appears in the west as usual, but a minute or two later, fades from view like a quenched spark as it's eclipsed by Earth's shadow.

As the biggest manmade object in space, the space station reflects much more light than smaller satellites, often outshining the planet Jupiter. Normally, it shines with a steady light, appearing brightest during overhead passes because it's closest then, only about 250 miles (400 km) directly above your nose. When it passes to the north or south of directly overhead or when near the horizon, it appears fainter because you're looking off in the distance to see it. The farther away something is, the fainter it appears.

Here are some fascinating things to watch for during a pass:

- Color: The station looks pale yellow because its solar arrays are insulated in a gold-colored foil called kapton.
- Brightness: When it crosses near the overhead point, the ISS is only about 250 miles (400 km) from you and absolutely brilliant. But when you first spot it low in the west, it's more like 870 miles (1,400 km) away because of the added horizontal distance between you and the craft, the reason it looks fainter. Like the Moon, the ISS goes through phases as the angle between the station, the observer and the Sun changes. When low in the west, the station's edge is illuminated by the Sun just like a crescent moon. As it ascends, more of the giant structure gets lit up by the Sun. When due south, it's roughly a "half-moon," and when it moves into the eastern sky, a waxing gibbous. The more it fills out, the brighter it gets . . . to a point. Even though the station's nearly "full" when it descends in the east, it's farther away, so it begins to fade. The ISS is brightest when it's high in the eastern sky but not directly overhead.
- Flares: Due to sunlight reflecting off the solar panels, you'll sometimes see a sudden surge in brightness before it fades back to normal.
- Sunset: When the ISS is eclipsed by Earth's shadow, the Sun sets from the point of view of the astronaut crew. On the ground, we see the station quickly turn from pale yellow to orange and then red. Binoculars will show this colorful transition—a mini must-see!

- Visits: Resupply tugs make regular runs to the space station and the Russian Soyuz spacecraft (at the time of this writing) ferries astronauts to and from the station. A cargo ship looks like a star, though not as bright, that tracks along with the space station until it docks.

There's talk about de-orbiting the space station sometime in the future, but Congress has extended NASA's operation of the ISS through 2024. What will happen beyond that date isn't certain, but there's discussion of transitioning the operation to private industry.

How to see the International Space Station

Now that you know *what* to look for, you'll just need the *when* and *where*. Because its orbit is steeply tilted, the ISS is visible across nearly every populated region of the planet. I've included several resources—websites and phone apps—so you'll have everything at your fingertips. If you're not familiar with where north, south, east and west are, you can either use your mobile phone—many have a built-in compass app—or just face the direction of sunset. That's west and the direction from which to anticipate the space station's arrival. Facing west, north is to your right, south to your left and east at your back.

Be aware that the space station has visibility "seasons." After several weeks of passes at dawn, it transitions to the evening sky for the next few weeks. At the end of the evening run, it's only out during the day (and invisible) for several weeks before transitioning back to dawn.

RESOURCES

- NASA's Spot the Station website. Sign up for email alerts: spotthestation.nasa.gov

- Heavens-Above. Login, pick your city and you'll get a list of the next 10 days of passes. Bonus—by clicking on a time link, a map pops up showing the station's path across the sky: heavens-above.com

- ISS Spotter for iPhone. A free app for your phone that includes pass times and alerts: itunes.apple.com/us/app/iss-spotter/id523486350?mt=8

- ISS Detector for Android: issdetector.com

- NASA space station FAQ: nasa.gov/centers/johnson/pdf/569954main_astronaut%20_FAQ.pdf

26

Iridium Flare

My fingers twitch as I write because there may be few Iridium satellites left to flare by the time you read this. The Iridium "constellation" of 66 telecommunications satellites (plus spares) was launched in the 1990s to provide worldwide wireless phone service. You wouldn't think anyone other than a service subscriber would care about these orbiting birds, but satellite watchers soon discovered that the satellites' highly reflective antennas had a penchant for reflecting sunlight down to lucky observers on the ground in brief dazzles of light called **flares**.

Iridium flares last several seconds and can appear up to 25 times as bright as the brightest planet, Venus. The satellites orbit at an altitude of about 483 miles (777 km), nearly twice as high as the space station, so they're typically faint and go unnoticed. But during a flare, the satellite quickly brightens, reaches greatest brilliance and then fades away, all in about 20 seconds. The brightest flares cast shadows—just remember to turn around and look instead of standing there mesmerized by the light like I usually do.

▲ *A bodacious magnitude -7.6 flare from the Iridium 3 satellite practically knocked me over when I took this picture. Both before and after the flare, the satellite was faint or invisible with the naked eye and left only a thread-like trail during the time exposure. Credit: Bob King*

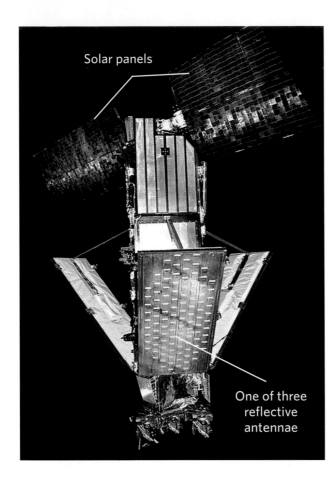

Solar panels

One of three
reflective
antennae

The brightest flares are so intense, they'll practically knock you over; they also make for great photos with this basic setup: camera, wide-angle or "normal" 50-mm lens, tripod and the ability to take a time exposure of at least 15 seconds. Using the directions in the How-To Section (page 105), find the time when a flare will occur at your location. Ten minutes in advance, set the camera lens to manual and camera exposure to 15 seconds or, better, 30 seconds and ISO 800. Open up the lens to its lowest f/stop, usually f/2.8 or f/4, to allow in the maximum amount of light. Carefully focus the camera on a bright star then point it to where the Iridium will appear. When you see the satellite begin to brighten, press the shutter button.

All this assumes a few Iridiums will still be in orbit in 2018 and beyond. The older Iridiums are being replaced by the new Iridium NEXT constellation with completion expected in the spring of 2018. Iridium NEXT uses an entirely different antenna design that doesn't produce bright flares.

Chances are there will still be a few of the golden oldies around for us to enjoy for a while, but don't hesitate or you'll miss out. These amazing sky bursts will someday come to an end.

Iridiums aren't the only satellites that flare. The final rocket stage that boosts a satellite into orbit sometimes goes along for the ride and becomes a satellite itself. Stages and some defunct satellites tumble and occasionally produce flashes but not with the predictability of an Iridium. Iridiums are so on schedule that you can impress an unsuspecting family member or friend with your predictive prowess. Tell them you have a gut feeling something's going to happen in the sky tonight, then step outside shortly before flare time, look in the satellite's direction and wait for the flare. When it happens as predicted, you'll be praised for your clairvoyant superpowers. Relish the moment, then tell them the real story.

▲ *Iridium satellites were built with three highly reflective antennae, the reason they reflect the sun so efficiently when lined up favorably with an observer on the ground. Source: Iridium Communications*

How to see an Iridium flare

My favorite website for flare predictions is Heavens-Above. Login and select your location and you're ready to start. On the home page under Satellites, click on Iridium Flares and you'll get a list of a week's worth of flares. The time is on the 24-hour clock, so 19:15 means 7:15 p.m. and 22:45 is 10:45 p.m.

Brightness is given on the magnitude scale, the same one astronomers used to rate star brightness. Without getting into detail, the higher the negative number, the brighter the flare. At minimum, flares are similar to the brightest stars in the sky, but at maximum (as high as –9), they're super-duper bright. Altitude is given in degrees. A balled fist held at arm's length against the sky spans about 10°.

The next set of columns in the table show the direction in which to look, the satellite's ID number, the distance to the center of the flare (location where it will appear most brilliant), its brightness at that spot and the Sun's altitude at the time of the flare (the numbers are all negative because the Sun has already set). For a map showing exactly where in the sky the flare will occur, click on the Time link.

Let's explore the *distance to flare center* concept. If you scroll to the bottom of the date/time page below the sky map, you'll see a smaller map with a red line on it. That's the flare centerline. Imagine it as the center or most brilliant part of the antenna's reflection moving along the surface of the Earth. Along this line, you'll see the brightest flare possible. If the red line is near your location, you're in good shape, but if your house is 50 km off the line, the outburst will be more modest. Some satellite watchers hit the road to see the brightest flares.

RESOURCES

- Heavens-Above: heavens-above.com

- Iridium NEXT: argo.ucsd.edu/sat_comm_AST13.pdf

- Sputnik! Free app for iPhone that provides both ISS and Iridium predictions for any location: itunes.apple.com/us/app/sputnik/id393001070?mt=8

- Iridium for Android: play.google.com/store/apps/details?id=com.androidsimple.iridium&hl=en

Vesta, Brightest Asteroid

Asteroids often make the news, especially online, where they seem intent on destroying humanity every couple of months. Robotic surveys like Pan-STARRS (University of Hawaii) and the Catalina Sky Survey (NASA and University of Arizona) constantly monitor the sky looking for potentially hazardous satellites that could impact the Earth. And every day, they discover new, Earth-approaching objects, most just a few dozen feet across. Astronomers follow up on the observations to determine their orbits and predict whether or not they might pose a future hazard.

More than 90 percent of near-Earth asteroids larger than one kilometer (0.6 mile) have been discovered to date, and none of them are making a beeline to Earth for at least the next 100 years, so go ahead and order dessert. You've got time.

On the other hand, we know for a fact that asteroids have struck Earth in the past, with devastating results. If you've never been to Meteor Crater in Arizona, located off I-40 37 miles (60 km) east of Flagstaff, put it on your list. At three-quarters of a mile (1,200 m) across and 560 feet (170 m) deep, it not only fills your field of vision but recalls the terror that must have ensued during its excavation some 50,000 years ago.

It didn't take much to make that big hole. An iron–nickel asteroid about 100 to 160 feet (30 to 50 m) across slammed into the ground at 26,000 mph (12 kmps) and exploded with a force 150 times greater than the bomb dropped on Hiroshima. The blast excavated 175 million tons (158 million tonnes) of rock, created winds in excess of 600 mph (1,000 kmh) and laid waste to mammoths, sloths and camels that roamed the land at the time.

To put this into perspective, meteoroids, which are itty-bitty asteroids, pepper the Earth and land as meteor*ites* or burn up and become dust flakes in the atmosphere all day, every day. Larger meteoroids, like the 65-foot (20-m)–wide boulder that exploded in the air 18.5 miles (30 km) over Chelyabinsk, Russia, on February 15, 2013, create shock waves that can damage structures on the ground, cause injuries from flying glass and drop significant meteorites. But even a rock this size rarely survives intact; it's traveling so fast that when it hits the lower atmosphere, it might as well be running into a brick wall. Most shatter into pieces, exactly what happened at Chelyabinsk. But if the pieces are large or if an object that size somehow survives intact, it can cause significant damage to a city.

NASA's current asteroid goal is to find 90 percent of all Earth-approaching asteroids larger than 460 feet (140 m) across, the ones that would cause regional destruction or a major tsunami.

Now that you have a healthy respect for asteroids, let's go out and look at one. Asteroids big and small make their home in the asteroid belt, a zone between Mars and Jupiter that contains between 1.1 and 1.9 million asteroids larger than one kilometer (0.6 mile) and billions of smaller ones. Although the total number is large, they're spread across a great volume of space, separated by an average distance of 600,000 miles (965,000 km). Unlike movie depictions, where they seem to swarm like honeybees, a spacecraft passing through the belt on a mission to the outer planets is quite safe.

Vesta, designated 4 Vesta because it was the fourth asteroid discovered, is the second largest after 1 Ceres with a diameter of 326 miles (525 km). It was discovered in 1807 and named for the goddess of home and hearth from Roman mythology. Vesta has a bright surface compared to many other asteroids, making it the only one that's occasionally visible with the naked eye. When closest to the Earth at opposition, it ranges from magnitude 5.3 to as faint as 6.6.

Because magnitude 6 stars are visible to the average person from a dark, country sky, Vesta is not difficult to find, even from outer suburbs, when shining brightest. Like everything in skywatching, you just have to know where to look. I've seen Vesta a few times without optical

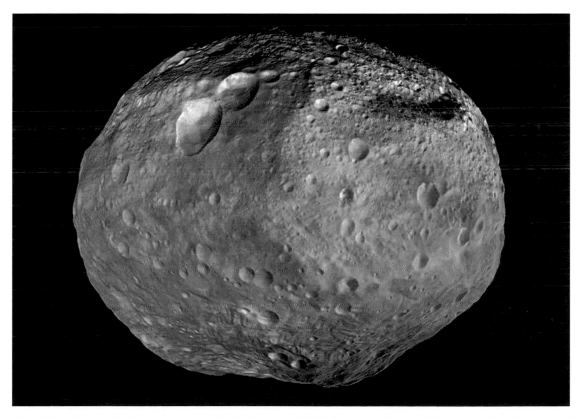

▲ *Vesta is one of the largest asteroids in the asteroid belt between Mars and Jupiter with a diameter of 326 miles (525 km); it's also the brightest asteroid visible from Earth. Source: NASA/JPL/MPS/DLR/IDA/ Björn Jónsson*

aid, and it looks exactly like a faint star. All asteroids look like stars in most telescopes, the reason they're called asteroids, from the Greek word "star-like."

Only the largest telescopes can clearly make out the shapes of the largest asteroids. To see them in all their rocky glory, a spaceship is much better. That's exactly what NASA did in 2007 when it launched the Dawn probe to Vesta for more than a year of in-orbit study. Dawn revealed that Vesta was more complicated than a simple ball of rock. It was a protoplanet.

A small asteroid begins its journey to protoplanethood when it grows large enough to collect a critical amount of radioactive debris during its formation. The material releases heat that melts the interior, causing heavy materials like iron to sink to the core and lighter rocks to rise and form a mantle and crust. At this point, astronomers say the object has "differentiated," the asteroid equivalent of getting a high school diploma. Both Earth and Vesta got their kindergarten diplomas about the same time, more than 4 billion years ago, and both celebrated the milestone with the release of vast flows of molten lava across their surfaces.

Vesta stopped there, but the much bigger Earth had more resources to work with, including plate tectonics and life, both of which continue to rework the surface and atmosphere of the planet to this day.

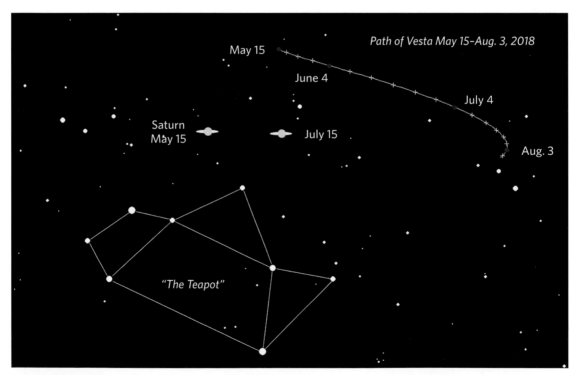

▲ *Use this map to find Vesta during its very favorable 2018 appearance, when it will be visible in binoculars and even with the naked eye from a reasonably dark sky. Saturn—its position shown from mid-May to mid-July—and the familiar "Teapot" of Sagittarius will be nearby and help in finding the asteroid. Vesta's position is marked every 5 days and stars are shown to magnitude 6. Source: Chris Marriott's SkyMap*

Vesta's brightness varies depending on its distance from the Sun. When closest to Earth, it's quite easy to see in binoculars, even from moderately light-polluted sites, and visible with the naked eye from rural skies. Try finding it first with binoculars then see if you can spot it with your eyes alone. Like a planet, Vesta orbits the Sun and inches across the sky from night to night. If you spot it one night, you'll be able to see it move when you return to look a couple nights later. Get to know Vesta—I promise this asteroid won't crash your party anytime soon.

How to see the asteroid 4 Vesta

The best times for viewing are at opposition. Those dates, the asteroid's distance from Earth and its brightness through the year 2025 are listed below. Vesta is well placed and easily visible for several months around the time of opposition, so you'll have plenty of opportunities to see it.

- June 19, 2018/106 million miles (170.6 million km)/magnitude 5.3. As bright as it gets! A wonderful time to spot your first asteroid.
- November 10, 2019/145 million miles (233.3 million km)/6.5
- March 9, 2021/126.5 million miles (203.6 million km)/6.0
- August 17, 2022/119 million miles (191.5 million km)/5.8
- December 22, 2023/147 million miles (236.6 million km)/6.4
- May 5, 2025/109.7 million miles (176.5 million km)/5.6

You'll generally need binoculars and of course a map showing where to look. I've included a locator map for 2018. Because asteroid orbits are subject to change, it's best to use a live source or star-mapping program after that time. Several are included in the resources section below.

RESOURCES

- TheSkyLive includes daily maps of Vesta's position and other bright asteroids: theskylive.com/vesta-tracker

- The Stellarium app will show lots of asteroids. Type in the name in the search feature, and you'll go directly to the asteroid of your choice. To update asteroid data, so it always shows the current, accurate position of Vesta, download and consult the User Guide at: sourceforge.net/projects/stellarium/files/Stellarium-user-guide/0.15.0-1/

- Asteroid Basics: solarsystem.nasa.gov/planets/asteroids/indepth

- Dawn Mission: dawn.jpl.nasa.gov/

28

Zodiacal Light

Spring feels good after winter. The air softens, the Sun sets later and, at least in my neighborhood, the frogs return to sing lusty songs all the night long. Evenings from late February to early May, when the Moon's not out, skywatchers can see one of the largest and yet subtlest sights in the solar system—the zodiacal light. The name hints at where we might look to see it, along the ecliptic highway traveled by the Sun, Moon and planets.

In early spring, that invisible highway is pitched at a steep angle to the western horizon. As twilight melts into night and the stars begin to show, look for a fat finger of softly glowing light tilted on its side in the western sky pointing toward the Pleiades star cluster. It runs straight up the ecliptic, so you'll occasionally see bright planets like Venus shining there. To the uninitiated, it strongly resembles yet another dome of light pollution from a nearby city, but the glow's distinctive

∧   *In a delightful coincidence, a meteor, likely a fragment of a comet, flashed across the dusty finger of* *zodiacal light on a mid-October morning. Credit: Bob King*

tapering shape and tilt set it apart.

What you're seeing is part of a vast cloud of fine dust left by passing comets and crashing asteroids aglow within the plane of the solar system. The mass of it extends from the Sun to at least Jupiter. The same way the Sun turns your breath into a brilliant cloud of vapor on a cold winter day, sunlight backscattered by interplanetary dust lights up as the zodiacal light.

The bottom of the zodiacal light cone is both broader and brighter because it's closer to the Sun and reflects sunlight more strongly. As you look up and along the cone, your gaze takes you farther from the Sun—the light source—and the dust dims and fades until merging with the sky background. Usually.

Under pristine skies, the cone tapers into an ultra-faint sash about 5° wide (three fingers held horizontally at arm's length) called the **zodiacal band** that circles around the sky to the opposite horizon, passing through the zodiac constellations. If you're fortunate enough to see the band, you're blessed with some of the darkest skies on the planet. I've never seen the full band, but stubs sticking out from the zodiacal "thumb" are visible on good nights.

The zodiacal light resembles the Milky Way in its smoky appearance and brightness. Dark skies are necessary for a good view. Just as you'd make a trip to the countryside for a good look at the Milky Way, you'll want to do the same for the zodiacal light.

Plan well as this unique phenomenon has but two viewing seasons for midnorthern latitude observers: late dusk in the western sky from mid-February through early May, and again from

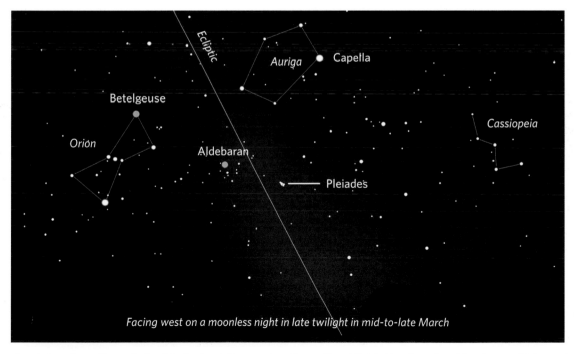

Facing west on a moonless night in late twilight in mid-to-late March

▲ *Late winter through April is the best time to view the zodiacal light. Head out in mid to late twilight and watch for the lowering cone of diffuse light to poke up from the western horizon as twilight fades to night. Source: Stellarium*

mid-September through November in the eastern sky before dawn. During these times, the ecliptic is most nearly vertical to the horizon, so the cone of light stands head and shoulders above the horizon haze. At other times of the year, the ecliptic meets the horizon at a shallow angle and the zodiacal light "lays low," blending in with the horizon glow.

Much of the dust eventually spirals into the Sun and gets replenished by new comets over time. The simple fact that we can still see the zodiacal light demonstrates that millions of comets have been coming from the farthest reaches of the solar system for brief visits to its warmer core since the planets and Sun formed some 4.5 billion years ago. One additional curiosity: While the zodiacal light looks atmospheric, it's most definitely not. If you were to remove Earth's atmosphere, it would still stand tall. If you're looking for the literal "sands of time," this is it.

How to see the zodiacal light

You'll need to avoid two things: light pollution and the Moon. For evening viewing in the spring, plan your trip to dark skies any time from two days past full moon until one day past new moon. That gives you almost two weeks to catch the sight each month.

Find a location with a wide-open view of the western sky and look from 80 minutes to 2.5 hours after sundown. The light stands tallest on the earlier end of that window. Sometimes, observers don't see the phenomenon because it's much bigger than imagined—easily six fists tall from the horizon and two to three fists wide. Take in the full sweep of the western sky, pivoting your head from left to right. You're looking for a broad pillar of what looks like fog, but not so thick or bright that it dims the stars.

In fall, plan your outing anytime from two days before new moon to one day before full moon. Find a location with a wide-open eastern exposure and look between 2.5 hours to 80 minutes before sunrise.

RESOURCES

- Moon phases calendar: moonconnection.com/moon_phases_calendar.phtml
- Light pollution map to help you find dark skies where you live: lightpollutionmap.info

29

Best Double Stars

There are so many double stars in the night sky, it's surprising the Sun isn't one. More than half of stars are born in pairs or multiples and live their lives with one or more companions. **Double stars** orbit around their common center of gravity. When each star is of similar size, that point is midway between them. When one is much more massive than its partner, the partner, called the **secondary**, orbits around the primary star.

Double stars are scattered like jewels across the entire sky with hundreds visible in small telescopes. If you were moved by the sight of the binary suns, Tatoo I and Tatoo II, on Luke Skywalker's planet, Tatooine, in the first *Star Wars* movie, wait until you see the real thing. When looking at double stars through the telescope, images from science fiction can help us picture what it might be like to see them up close from an orbiting planet.

I've heard telescopes sometimes referred to as "star-splitters" because they can divide stars that appear single to the human eye into two. With our ax of finely ground glass, let's take a look at several of the brightest, most colorful and charming double stars across the seasons. All the

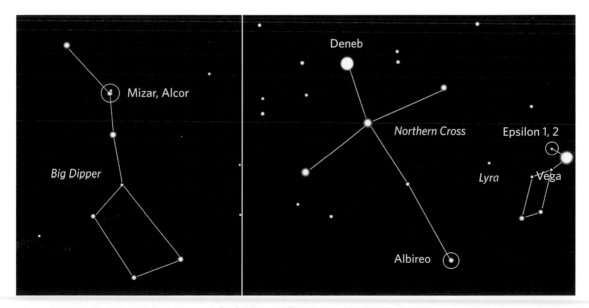

▲ *Use these simple star maps to find Mizar, Albireo and Epsilon 1, 2 Lyrae in your telescope. Source: Stellarium*

stars we'll visit are true doubles, where each sun revolves around the other as they travel together through space. Some stars *appear* double but are actually unrelated stars seen along the same line of sight; astronomers call them **optical doubles**.

When it comes to fuzzy stuff like galaxies or nebulae, it helps to learn about your subject beforehand to truly appreciate what you're looking at. But double stars are pure eye candy, and while it doesn't hurt to know more about them, too, it's easy to enjoy these stellar gems as shiny objects.

Albireo—visible in spring, summer and early fall

If you seek only one double star on our must-see list, make it this one. Albireo shines at magnitude 3 at the foot of the Northern Cross, so it's easy to find with the naked eye from the time it climbs up the eastern sky in May all the way through November, when the Cross stands straight up and down in the northwestern sky. Through a telescope magnifying 30x, Albireo resolves into a fiery orange star with a pale-blue companion set against a spectacular backdrop of fainter Milky Way stars.

The warm-hued primary star is an **orange giant** star 950 times brighter than the Sun and 50 times as big; the fainter star is hotter than the primary and 190 times brighter than the Sun. Albireo's stars are so far apart that it takes more than 75,000 years for them to complete an orbit about one another. Because the pair lies about 430 light-years from Earth, the light you see left this duo about the same time Shakespeare wrote his first plays and 20 years before the telescope was invented.

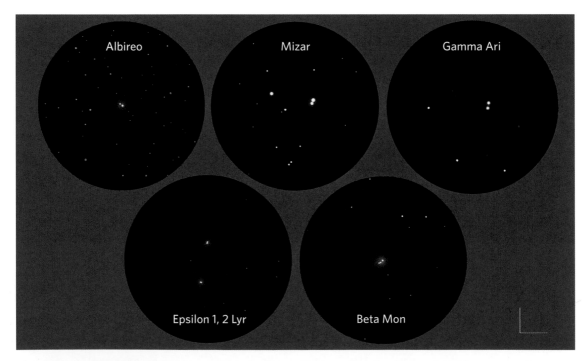

➤ *Double stars are the sky's gems and jewels. These sketches of our five featured doubles accurately capture their delicate hues and fragility when viewed through a telescope. Credit: Jeremy Perez*

Mizar—visible in winter, spring, summer and early fall

Easy to find in the bend of the Big Dipper's handle, Mizar is really a sextuple star, two of which are visible without optical aid and the third in any telescope magnifying just 30x. The brightest companion, named Alcor, is faintly visible just above Mizar. Play your eye around the star with averted vision, and you should be able to see it. The pair's also called the "**horse and rider**," a fitting image. In ancient times, before the invention of eyeglasses, the double was considered a test of good vision.

Magic happens when you aim a telescope at the pair. Mizar cleaves into two closer-set jewels with Alcor off to one side.

Mizar was the first double star discovered with a telescope (1617), only a few years after its invention. In the nineteenth, twentieth and twenty-first centuries, astronomers continued to study Mizar. Using specialized instruments, they discovered that each of the three was accompanied by a close companion, making for six stars in all!

You can look up this most interesting multiple star almost any night of the year because the Big Dipper and Mizar are circumpolar for observers at midnorthern latitudes: They circle around the North Star and stay above the horizon all night long.

Epsilon Lyrae aka The Double Double—visible in spring, summer and fall

Like Albireo, the Double Double is a must-see, and since both are in the same part of the sky, you can look at one then move on to the other. Epsilon isn't particularly bright at magnitude 4.6, but it sits practically on top of Vega, the sky's fifth brightest star and the brightest member of the Summer Triangle, so it's easier to find than you'd think. With the naked eye, most of us see only a faint pinpoint, but if you look closely, alternating between direct and averted vision, you might just be able to see two stars here, Epsilon-1 and Epsilon-2. Any pair of binoculars will show them. But a small telescope capable of magnifying 100x to 150x will reveal that each of the two is double again!

Gamma Arietis or Mesarthim—visible in fall and winter

In most double stars, one sun is brighter and the other fainter, but some pairs have equally bright stars. These so-called **equal doubles** make for a special visual treat, and the most famous of them is Gamma in the constellation Aries the Ram. You'll find this small, hockey-stick–shaped constellation just to the east of the much larger asterism the Great Square of Pegasus. Gamma, also known as Mesarthim, marks the end of the stick and is easily visible with the naked eye at magnitude 3.8.

Through a small telescope at 30x, it might still appear single, but turn up the magnification to 50x or higher and you'll see what looks like a pair of car headlights shining back. Both Gamma-1 and Gamma-2 are about 50 times brighter than the Sun and nearly three times as massive. Their actual separation is more than a dozen times Pluto's distance from the Sun or at least 47 billion miles (75.6 billion km). They sure seem closer to your eye but only because they're so far away—about 164 light-years.

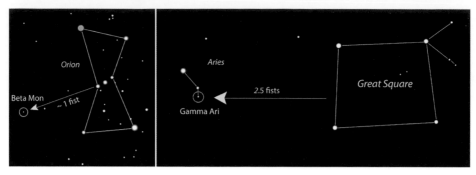

▲ *This set of maps will guide you to Beta Monocerotis and Gamma Arietis.*
Source: Stellarium

Because the stars in each pair are very close to each other—they resemble small ants—to see them clearly requires "steady seeing," another name for calm atmospheric conditions. That's hit and miss, so you might have to try on several nights until the air is tranquil.

For sharp views at higher magnification, a telescope should be about the same temperature as the outdoor air. If you take your scope from a warm house into a cold night and observe Epsilon first thing, heat radiating from the tube will soften and blur the view, making it impossible to split the two tight pairs. Remember to let your scope cool down first.

All four stars in the quadruple are larger and brighter than the Sun and 162 light-years from Earth. The stars of Epsilon-1 orbit around each other about every 1,800 years and come as close to each other as 73 times the Earth–Sun distance. The Epsilon-2 pair takes 724 years for an orbit with a minimum separation of 95 times the Earth–Sun distance.

While humans won't be landing on planets orbiting these or any other double stars in the foreseeable future, we can still imagine our way there with the help of a small telescope on the next clear night.

Beta Monocerotis—visible in winter and early spring

Beta Mon (for short) shines from the modern constellation of Monoceros the Unicorn just east of Orion. Although only magnitude 4.6, it's a hidden treasure. At 40x magnification, two stars appear, but if you go up to 75x or higher, the eastern star splits into a close pair, making for a stunning triple star. All three are hot, blue-white stars 1,300 to 3,200 times brighter than the Sun and about 700 light-years away.

How to see double and multiple stars

A small telescope and low magnification eyepiece are all you need for the first three pairs, but the Double Double requires higher magnification, around 150x, while Beta Mon is best at 75x and up. Use the provided maps or a phone app in the suggested seasons to locate each of them, then aim and look.

Ring Nebula

I recall a summer night years ago with the telescope set up on the driveway. As I poked around the night sky, my younger daughter, Maria, stepped out of the house. "Whatcha lookin' at?" she asked. "The Ring Nebula," I answered. "Wanna see it?" She climbed up the wobbly stepladder, looked into the eyepiece and saw nothing less than the end of the Sun as we know it.

Like Saturn, the Ring Nebula is marvelous at first sight. Easy to find midway between the two bottom stars that outline the harp in the constellation Lyra, the Ring looks like a smoke ring of soft, nebulous light afloat in a field of stars. You can't put it on your finger, but it does fit nicely in the eye.

The Ring's shape and crisp outline reminded an earlier generation of astronomers of a planet, hence its designation as a **planetary nebula**. Its other name is M57, for the 57th entry in a catalog of 110 deep-sky objects compiled by the eighteenth century French comet hunter Charles Messier.

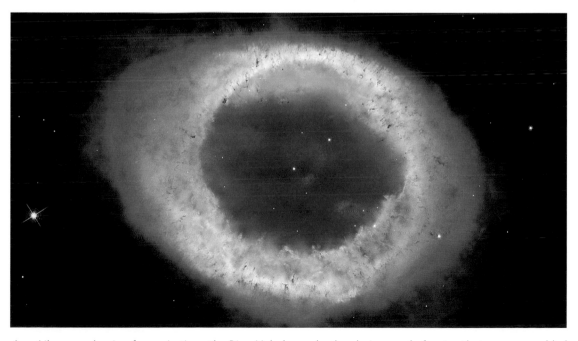

▲ *Like a smoke ring frozen in time, the Ring Nebula marks the glorious end of a star that once resembled the Sun. Source: NASA, ESA and The Hubble Heritage/Hubble Collaboration*

During his nightly comet quests, he'd occasionally run across fuzzy objects—nebulae, galaxies and star clusters—that masqueraded as his favorite prey. Messier didn't have the benefit of current-day maps showing the positions of literally thousands of deep-sky objects, so when he stumbled onto something fuzzy, he had to patiently observe the object or return to it another night to see if it moved before deciding its nature. To avoid being waylaid by every fuzzy whatsis, Messier compiled a list with positions and descriptions of what he called his "embarrassing objects."

On the morning of January 31, 1779, while making observations of Bode's Comet, Messier discovered what would come to be called the Ring Nebula with his small refracting telescope, noted in *Histoire de l'Académie Royale des Sciences,* 1779: "It is very dim but with a sharp boundary; it is as large as Jupiter and resembles a planet which is fading away."

One of his contemporaries, Antoine Darquier, was also searching for the comet with a 2.5-inch (63-mm) refracting telescope and stumbled across the Ring Nebula later that February. Here's his description from his letter to Messier that appeared in the *Histoire de l'Académie Royale des Sciences*: "A very dull nebula, but perfectly outlined; as large as Jupiter and looks like a fading planet."

Despite appearances, the Ring has no relation to the planets. It's a shell of gas about a light-year across centered on a tiny, hot, super-dense star called a **white dwarf**.

A white dwarf is what's left after a sun-like star runs out of nuclear fuel. Deep in its core, under tremendous heat and pressure, the star fuses hydrogen into helium to make energy. When the hydrogen runs out, the core shrinks and heats up even more, igniting helium which burns to form carbon and oxygen. During this brief phase of the star's life, its outer atmosphere balloons out to hundreds of times its current size, and the star becomes a red giant. When the Sun swells into a giant in about 5 billion years, astronomers predict it may become large enough to engulf the Earth; Mercury and Venus will be toast.

Once the helium fuel is exhausted, nuclear fusion stops. But the core continues to collapse and heat up, ultimately becoming a white dwarf, a star with the mass of the Sun compacted into a sphere the size of Earth. Gusty "winds" blowing from the star combined with the pressure of light

▲   *Ring Nebula map—Located about midway between the bottom two stars in the summer constellation Lyra the Harp, the Ring is easy to find. Source: Stellarium*

streaming from its core expels its outer gaseous envelope into space like a child blowing a bubble. Powerful ultraviolet radiation from the dwarf excites the expanding shell to glow, and a planetary nebula is born. The Ring Nebula's star made this transition some 20,000 years ago when humans still hunted woolly mammoths.

Messier would kill to look through one of today's telescopes. The optics are sharper, and they gather more light for their size, so even a 3-inch (75-mm) scope at a magnification of 50x and higher will show the Ring as a puffball of light with a slightly darker center, the whole resembling a cosmic donut. A 6-inch (150-mm) or larger telescope will show that it's more oval than circular and more clearly show the darker center. I've only seen the central white dwarf with 11-inch (280-mm) and larger telescopes on nights of excellent seeing. Then, it appears as the tiniest flicker of light, hovering at the limit of vision.

When you look at the Ring through a telescope, you walk a line separating past from future. In one direction, we join Messier in his moment of discovery on that cold January morning more than 200 years ago. In the other, we see like seers to a time when the Sun will be transformed into an unrecognizable, planet-size star ringed by the shell of its former self.

How to see the Ring Nebula

The best times of year for viewing the Ring are summer and fall, when it's highest in the southern sky. First find Vega, the brightest star in the Summer Triangle, and use it to work your way around the outline of the Harp. The nebula sits between Gamma and Beta Lyrae, the two stars at the bottom of the Harp. Point the telescope between them and a little closer to Beta, then look through the eyepiece for a small ball of fuzz that resembles an out-of-focus star.

Lyra hangs out near the zenith at the times listed below, but it's up in the east all spring and in the west during the fall months.

- Early June around 2 a.m.
- Early July around midnight
- Early August around 10 p.m.
- Early September around 9 p.m.

Although visible in 50 mm and larger binoculars, it will look like a star. To appreciate its character, I recommend at least a 3-inch (75-mm) telescope and preferably a 6- to 8-inch (150- to 200-mm) scope. It looks like a small, gray dot at 30x. Higher magnifications upwards of 50x are better.

RESOURCES
- Star Chart app for iPhone and Android. This and other apps not only show the constellations but also plot the positions of the objects in Messier's catalog.

31

Perseus Double Cluster

Buy one bag of chips, get another one free! I always fall for that one. I'll come home and tell my wife, "Hey, I got a great deal." Right. The sky has plenty of 2-for-1 specials and your cost is always zero dollars unless you count mosquito bites and occasional cold fingers. But even that can be avoided by choosing the right time of year to "shop."

Just below the W of Cassiopeia, not far from the Andromeda Galaxy, where the band of the Milky Way gets skinny in Perseus, a hazy knot of light glows brighter than its surroundings. Although visible as a fuzzy patch with the naked eye from the outer suburbs and countryside, a pair of 40 mm or larger binoculars under moonless skies reveals not one but *two* side-by-side star clusters rich with pinpoint suns.

▲   *Just as we used Cassiopeia to find the Andromeda Galaxy (page 29), we can press it into duty again to point us to the Double Cluster. Credit: Bob King*

Welcome to the Double Cluster in Perseus. The two are aligned approximately east–west in the sky and bear the catalog numbers NGC 884 (eastern) and NGC 869 (western). NGC stands for New General Catalog, a compendium of 7,840 galaxies, nebulae and star clusters based on the *General Catalog* compiled by astronomer John Herschel in the nineteenth century. It grew out of the original *Catalogue of Nebulae and Clusters of Stars* created by his father, William Herschel, the eighteenth-century astronomer who discovered the planet Uranus.

The two clusters aren't a line-of-sight happenstance; they're actually near one another in space—about 300 light-years apart—and reside in an enormous cloud of gas and young stars called the Perseus OB1 Association, located in an outer spiral arm of the Milky Way Galaxy some 7,500 light-years from Earth. Take a minute to "sweep" the sky in the vicinity of the Double Cluster with those binoculars. You're sure to run across other starry knots and sprays. Where Perseus and Cassiopeia meet, star clusters abound like candy tossed at a parade.

Through a modest-size scope, the Double Cluster practically explodes with stars, a few of which shine vividly red. I don't care how many times you see it, I guarantee you'll return every fall for another look.

I'll never forget my first encounter with the Double Cluster as a 12-year-old. From my suburban Chicago neighborhood through a 6-inch (150-mm) reflecting telescope, I was stunned by what seemed like hundreds of stars. Several ruby-red ones added extra pizzazz to the scene, while a

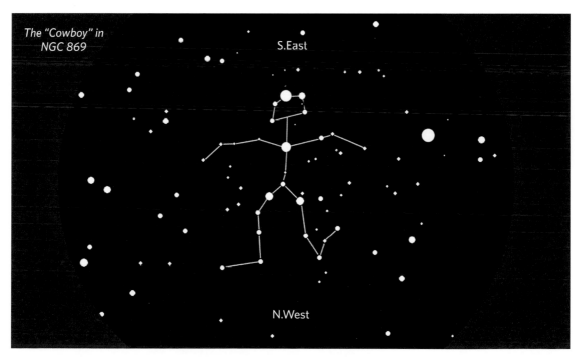

The "Cowboy" in NGC 869

S.East

N.West

▲　*Wonderful starry patterns big and small abound in the night sky including this "cowboy" inside NGC 869. Can you see it? The cowboy's head also does double duty as the cluster's "smiley face." Credit: Bob King*

group of suns shaped like a smile in the center of NGC 869 literally made me smile. What a tiny, beautiful asterism—be sure to look for it. And if you think you need lots of magnification to see things in the sky, these twin clusters will prove you wrong. 30 to 70x can't be beat.

Since that first experience, I make sure to drop by for a look whenever Perseus stands high in the northeastern sky. Thanks to the cluster's relative proximity to the North Star, it's visible from mid-August through early April.

All star clusters are gravitationally bound collections of stars born from nebulae. Astronomically speaking, the twin clusters began life recently: only six million years ago for NGC 869 and fourteen million for NGC 884. In comparison, the Sun's an old fuddy-duddy, having recently celebrated its five billionth birthday, making it some 500 times older.

The Double Cluster is loaded with freshly minted supergiant and giant stars that shine with a fierce brilliance. That's why this duo is so bright and easy to find, the same reason it was first recorded more than 2,100 years ago by the ancient Greek astronomer Hipparchus.

How to find the Double Cluster

Whenever the W of Cassiopeia is easy to see in the northern sky is a good time to look for the Double Cluster. Use the map to guide you to this stellar treasure. A 3-inch (76 mm) telescope magnifying 25 to 30x provides a stunning view. The larger the telescope, the more stars you'll see and the more intensely bright they'll appear. However, large scopes generally can't get down to the low magnification needed to see *both* clusters at the same time, so there's a trade-off. Don't forget binoculars. The view is quite good in a 50 mm pair, but large 15x70s or 15x80s mounted on a tripod will offer a sweeping vista of the two clusters hovering in 3D against draperies of stars.

RESOURCES

- *The Complex Lives of Star Clusters* by David Stevenson: amazon.com

Great Globular in Hercules

With the word "great" in its name, this cluster must really be something. Rest assured, you won't be disappointed. Also known as M13, the thirteenth entry in Messier's catalog, it's visible without optical aid on moonless spring and summer nights from the countryside as a small, fuzzy patch. Even *my* old eyes can still spot this magnitude 5.8 smidge of mist along the western side of the Hercules "Keystone" asterism.

If "fuzzy patch" seems to describe so many objects in the heavens, I understand your skepticism. We can't help this. Most everything in the universe is so far away, much of it looks the same at first blush. Remove stars to a great distance, and they merge into haze. That's why we use telescopes to expose these downy blobs for what they really are—stunning star clusters, mind-bogglingly distant galaxies and colorful nebulae. That's also why both amateur and professional astronomers desire

▲ *It's crowded in here! The Great Globular in Hercules jams some 300,000 stars into a sphere about 145 light-years in diameter. The view through a telescope will knock your socks off. Credit: Adam Block/Mount Lemmon SkyCenter/University of Arizona*

ever-larger telescopes. Bigger scopes deliver more light and turn fuzz back into suns. Let's hope my wife will appreciate this argument when it's time for that new telescope.

Based on photos taken with larger instruments, astronomers estimate M13 packs some 300,000 stars into a roughly spherical volume 145 light-years wide. When viewed through a telescope, the stars appear stationary as if painted on black canvas, but if we could speed up time so that for every minute a million years would pass, the cluster would palpitate with life, its stars moving randomly about but held together by the immense gravity of the whole.

Globulars are densely packed, generally spherical star clusters containing from 10,000 up to a million stars. They orbit the central bulge of the Milky Way Galaxy in a spherical halo thought to trace the limits of our galaxy in its youth when it was in the process of collapsing from smaller clouds of dust and gas. As the collapse proceeded, the Milky Way's rotation sped up for the same reason a skater pulling in her arms spins faster, causing the newborn galaxy to whirl itself into the flattened disk we know today.

And the globulars? They remain as border guards along the periphery of a great stellar empire. Because the clusters formed early in the galaxy's evolution, their stars are ancient, 10 billion years and older. When you look at M13, you're seeing a remnant from the birth of the Milky Way before the Earth was even so much as a possibility.

We know of about 150 globulars in the Milky Way Galaxy. Some galaxies have many more; roughly 12,000 orbit the bright galaxy M87 in Virgo and more than 500 have been discovered in the nearby Andromeda Galaxy. The Hercules cluster is one of the brightest and richest, the reason you simply *must* see it. Binoculars begin to show some structure—a brighter core, where the stars are

∧ *This is my crude attempt to imagine what the night sky might look like if Earth were at the center of a globular cluster. Credit: Bob King / Stellarium / Background image by the Hubble Space Telescope / NASA / ESA*

more concentrated, surrounded by a fainter halo, where the stars thin out. But to really appreciate this stellar beehive, you'll need at least a 6-inch (150-mm) telescope. Then you can resolve (split apart) M13's outer halo into hundreds of individual stars and even see starry sprinkles across the denser core. Use a magnification of around 100x to 200x for the clearest view.

I love listening to the reaction of someone who sees the cluster for the first time through a larger telescope (8- to 12-inches [200- to 300-mm]) under a dark sky. While the visual and emotional impact of a planet is immediate, globular clusters have this delicious, built-in delay between recognizing what you're looking at and then truly seeing it. Before the realization sets in, there's a moment of silence, followed a second later by something along the lines of, "Oh my God, I can't believe all those STARS!" In these larger instruments, you can resolve most of the cluster and see one of its most striking features: tendrils of stars that extend from the core through the

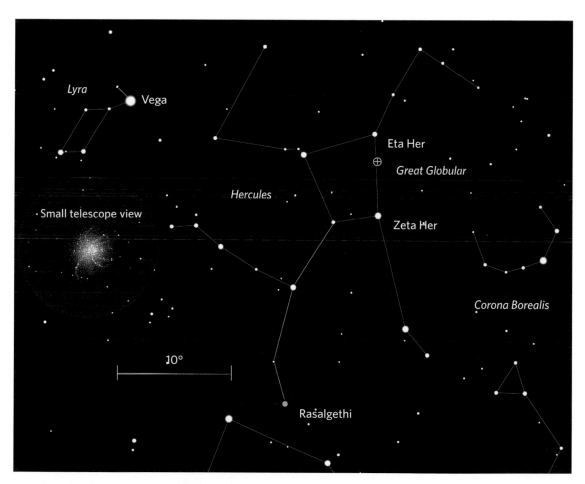

▲ *The globular hides out in the hinterlands of Hercules about two fists to the west of Vega, the brightest star in the Summer Triangle. Use Vega to spot the four-sided asterism called the Keystone then point your telescope at a spot one-third the way down the asterism's right side to find the cluster. Source: Stellarium*

125

halo-like crab legs. If you can't afford a bigger scope, check to see if there's an astronomy club in the area. They'd be happy to share the sight.

If seeing all those stars has you wondering what the night sky would look like from a planet orbiting a star *inside* a globular cluster, let's check it out. The average density of stars in a typical globular is about one per light-year, but in the standing-room-only core, it's closer to one star *per solar system* diameter. From deep within M13, you'd stare into a sky spangled with thousands of stars many times brighter than the brightest stars we see from Earth. They'd throw more light than a full moon, making finding your way at night a simple task and cast multiple, overlapping shadows.

On November 16, 1974, Frank Drake, then professor of astronomy at Cornell University, used the Arecibo radio telescope in Puerto Rico to beam a digital "postcard" to the Hercules cluster with basic information about us humans. It included our physical appearance, numbering system, the composition of DNA and a graphic of the solar system, showing where the message came from.

Consisting of 1,679 binary bits representing ones and zeros, it was our *first* deliberate effort in the modern-day era to toss a "message in a bottle" into outer space with the hope that an alien civilization might pluck it. And what better place to send it to than a rich, evolved star cluster that could potentially harbor tens of thousands of planets.

The missive, still barely "out the door," will have traveled 44 light-years by 2018, only 0.002 percent of the way to the great cluster. But humming along at the speed of light, we appreciate the spirit of hope that propelled it into the unknown.

How to see the Hercules Globular Cluster

The best time of year to view the cluster is when it's well-placed in the evening sky in the spring and summer months. M13 is located a little more than a third of the way from the brilliant Vega to equally bright Arcturus. The map will help you find it. Once you're in the right spot, see if you can pick up M13 with the naked eye using averted vision—it makes for a nice challenge. Then use binoculars or a telescope to go exploring. You can see the globular in any telescope, but the larger the better. Globulars come alive in big amateur scopes!

RESOURCES

- Stellarium or a phone app like Sky Chart described on page 10
- A telescope or access to a telescope through an astronomy club

33

Slender Moons

Some people salivate over vintage cars. For me, it's super-thin lunar crescents. More power to us both. My personal record for the youngest or thinnest crescent moon is 21.5 hours, and it was surprisingly easy to see. Nothing quite compares to the delicacy of an hours-old moon. Spider silk, maybe?

Wiry moons magically blend two dissimilar shapes into one: crescentic curve and two sharp points. Crescent moons inhabit the western sky after sundown and the eastern sky before sunrise. Since we see them in the context of the most colorful times of the day, they often take on the pink and orange hues of the hour, which adds to their charm.

Crescents, especially the thinnest, hover near the horizon and are only in view for a short time. Like an all-too-brief visit from a good friend, it's hard to see the Moon go. Because the scene may

▲ *A one-day-old moon is a delicate and stirring sight. Only the bottom edge is illuminated by the sun; the remainder is lit by sunlight reflecting off the Earth or earthlight (see page 161). Credit: Bob King*

not repeat itself for some time, we're reminded how temporary so many things in life are. On the other hand, the sight of a fresh-faced lunar crescent reborn at dusk once a month may have the opposite effect and inspire you with new purpose. Such is the wonder of nature that meaning is there for the taking.

While spotting a day-old moon isn't too difficult, anything under 20 hours (or 20 hours *before* new moon if you're seeking morning crescents) takes careful planning. Every hour younger means the Moon is that much thinner and closer to the horizon, making it more difficult to see. The record for the youngest moon seen with the *naked eye* goes to writer and amateur astronomer Steven James O'Meara, who nabbed a 15 hour 32 minute crescent in May 1990. Mohsen G. Mirsaeed of Tehran broke the record for the youngest moon ever seen *with* optical aid on September 7, 2002. He observed it from a mountain site using giant 40×150 binoculars and kept the sliver crescent in view for one minute. The Moon was just 11 hours 40 minutes past new and 7.5° from the Sun.

Can we do better? Probably not in terms of seeing the Moon with our eyeballs, but the ultimate record was set on July 8, 2013 by French astrophotographer Thierry Legault, who photographed the very moment of new moon. It happened smack in the middle of the day with the Sun up in a blue sky. The Moon lay south of the Sun at the time, so its extreme northern edge caught enough sunshine to show as the thinnest of crescents. He never saw it with his eye—only the camera recorded the historical moment.

▲   *Using a special telescope setup, the French amateur astronomer achieved the nearly impossible feat of photographing the new moon crescent on July 8, 2013. Credit: Thierry Legault*

Believe it or not, we can repeat this feat. During a partial or total solar eclipse, the new moon partially or completely blocks the Sun. The black circle we see is the new moon, its nearside completely in shadow. We'll cheat a little and call it a crescent s-o-o-o thin, you can't even see it.

Spring evenings and fall mornings are the best times for crescent hunting. That's when the ecliptic, the Moon's highway through the sky, stands at a steep angle to the western horizon (spring) or eastern horizon (fall). Within 20 to 24 hours of new moon, the new crescent climbs straight up into the sky and becomes easy to spot. In fall, the ecliptic meets the western horizon at a *shallow* angle and the Moon barely clears the distant trees.

Once you know the best viewing times (see next page), the rest I hope will be easy—a cloud-and haze-free sky with a wide-open view to the west for evenings and east for mornings. Oh, and bring binoculars. If the Moon proves elusive with the naked eye, binoculars will help pull it in. Start looking about 20 to 30 minutes after sunset or before sunrise for those 20- to 24-hour-old crescents.

As you study the Moon in binoculars, you'll notice right away that the skinny arc isn't smooth but ragged or broken into segments. These seeming breaks are caused by oblique lighting on crater walls and mountain peaks that create shadows long enough to bite into and hide portions of the Moon's limb or edge. With extremely thin moons, the crescent is shorter than normal and has a bended appearance.

You might also notice pale earthshine illuminating the remainder of the Moon. We'll explore this in more detail in a separate entry, but know that what you're seeing is sunlight reflected twice: once from the Earth to the Moon and a second time from the Moon back to your eyes. Because it's reflected sunlight and not direct, it only faintly illuminates the non-sunlit portion of the Moon.

If you're lucky, a bright planet—usually Venus—will be near the sunlit sliver and enhance the beauty of the scene. The Moon in twilight practically shouts, "Take my picture!" Sometimes you can handhold the camera because modern lenses come with built-in stabilization. If you can't, then mount the camera on a tripod and compose a scene with the Moon.

Next, open the lens wide, anywhere from f/2.8 to f/4.5, set your ISO to 400 or 800 and experiment with exposures from one-tenth to several seconds long. Check the back display and adjust the exposure as needed. Because the crescent presents a sharp, bright shape, your camera's auto focus should work fine.

If 24-hour-old crescents prove too taxing, 48-hour ones are much easier to see and still thin enough to thrill. They're also visible for a longer time and shine brighter in a darker sky. Keep in mind that the Moon's exact age depends on your time zone.

Example: If new moon occurs at 8 p.m. Eastern Time on March 15, the Moon will be 24 hours old at 8 p.m. the following evening, a perfect time to go out for a look.

How old will the Moon be on the same evening in the other time zones? 8 p.m. Eastern Time is the same as 7 p.m. Central, so Minnesotans must wait an additional hour for the Sun to set, at which time, the Moon will be 24 + 1 or 25 hours old. Coloradans see a 24 + 2 or 26-hour moon, while Oregonians see a 27-hour moon.

The key to figuring out the Moon's exact age is to consult an online calendar to find the time of new moon, then add hours (or days) until you arrive at the time you want to look for the thinnest crescent. So . . . if new moon happens at 6 a.m. on April 10 and you plan to observe it on the 11th at 9 p.m., the Moon at that time will be 1 day and 15 hours old.

I wish you many dusks and dawns under the sickle moon.

How to see a thin crescent moon

Use the resources below to identify the time of new moon and plan to catch your thinnest crescents on late winter and spring evenings and late summer and fall mornings. Every month presents an opportunity to see these young and old moons. Several are listed below. For future years, consult the resources section.

Dusk 2018–2019—Look 45 minutes to an hour after sunset low in the western sky

- June 14-15, 2018
- July 13-14, 2018—On the 13th for the United States, the Moon's age will vary from just 22 hours on the East Coast to 25 hours on the West
- December 8-9, 2018—Saturn to the left of the Moon on the 8th
- March 7-8, 2019
- April 6-7, 2019
- May 6-7—Just below Mars on the 7th

Dawn 2018–2019—Look 45 minutes to an hour before sunrise low in the eastern sky

- July 10-11, 2018
- August 9-10, 2018—On the 10th for the United States, the Moon's age will vary from 23 to 26 hours old
- November 5-6, 2018—Near Venus both mornings
- August 28-29, 2019
- September 26-27, 2019
- October 25-26, 2019—Just above Mars on the 26th

RESOURCES

- Abrams Sky Calendar. An easy-to-use, illustrated monthly calendar that tracks the motions of the Moon and planets and alerts skywatchers to slender moon opportunities. Available by mail subscription for $12/month: abramsplanetarium.org/skycalendar

- Moon phases/lunar calendar. Phase times are shown for your time zone and account for Daylight Saving Time. Very handy! timeanddate.com/moon/phases/

- Stellarium or a phone app like Sky Chart described on page 10 will show the Moon's phase and position any time of year.

34

Supermoons

The Moon is full every month, but not all full moons are equal in distance, size and brightness. If the Moon's orbit around the Earth were a perfect circle, they would be. But the Moon's orbit is an imperfect or "squashed" circle called an **ellipse**. At one end of the ellipse, called **perigee**, the Moon is closest to the Earth. At the other end, called **apogee**, farthest. It passes the perigee and apogee points once every 27.5 days, the time it takes the Moon to orbit the Earth.

While the Moon's average distance from Earth is around 239,000 miles (384,600 km), at perigee it is more like 225,000 miles (362,000 km). A perigee moon appears both bigger and brighter than average. At apogee, the Moon backs off to about 252,000 miles (405,500 km) and looks smaller and fainter.

A supermoon occurs when full Moon happens at or near perigee. A typical supermoon is around 7 percent bigger and 16 percent brighter than an average full moon, but during an unusually close perigee, the Moon can be 12 to 14 percent larger and 30 percent brighter than an apogee full moon.

Moon Perigee vs. Apogee

Perigree
218,500 miles away
Jan. 9, 2009

Apogee
251,000 miles away
May 20, 2008

⌃ *Over the course of its orbit, the Moon's apparent size varies because of it changing distance from Earth. A supermoon occurs when the full moon is at or near perigee, its closest point to Earth. Credit: James Schaff*

The term supermoon is a recent invention. It came into common use in 2011, but appears to have been coined back in 1979 by Richard Nolle, who describes himself as a certified, professional astrologer. His definition of a supermoon in his book *Horoscope* was generous: " . . . a new or full moon which occurs with the Moon at or near (within 90 percent of) its closest approach to Earth in a given orbit."

The opposite of a supermoon is an apogee moon also called a **mini-** or **micromoon**. No one celebrates these smaller than average moons. Like TVs, most of us prefer our full moons large.

With one and occasionally two full moons a month, it's not surprising that we get at least a few supermoons every year. In 2018, there were two, both in January. Because the second full moon in the month is nicknamed a Blue Moon, lucky skywatchers witnessed a Super-Blue Moon at month's end that year.

Some supermoons are more super than others. If the full moon occurs within a few hours of perigee, that supermoon will be a bit closer, bigger and brighter than a supermoon occurring plus or minus a day of perigee. Other factors are also at play that draw the Moon in closer yet.

When a full moon arrives at its perigee point, it makes a neat lineup with the Earth and Sun, both of which are gravitational heavyweights. The Sun in particular holds sway over the Moon; despite its much greater distance, it's far more massive than the Earth with a gravitational grip about twice as strong.

∧   *Supermoon measuring device—Make a simple moon-measuring device by cutting slots of different widths in an index card and finding one that "fits" the Moon. While not hard science, you can use it to record the difference in size between apogee and perigee moons. Credit: Bob King*

The neat alignment of the three bodies shifts the Moon's orbit slightly in the Earth–Sun direction, moving the perigee point a bit closer to Earth. But we're still not finished. If a supermoon occurs during late fall through midwinter, when Earth is closest to the Sun, that extra bit of solar tug will bring the moon in even *closer*.

All these things conspired to bring the November 2016 full moon just 221,524 miles (326,508 km) from Earth or about 4,000 miles (6,400 km) closer—equal to half Earth's diameter—than a typical perigee. Many of us at the time couldn't resist calling it the super-duper supermoon. 2016 was the last exceptionally close perigee; the next occurs on November 25, 2034.

While differences in the Moon's distance are significant, can you really tell if the Moon looks larger without a measuring device? After all, we can't drag a minimoon and plunk it next to a supermoon. We either have to recall the average appearance of a full moon and compare in our mind's eye or find a way to measure the difference. Some observers have excellent recollection abilities and can clearly see the difference. Others, like me, require help.

That's why I cooked up a little device I call the Supermoon Sighter that provides at least a rough measure of the Moon's apparent size using just the naked eye. With a pair of scissors, cut a series of slots of varying widths in an index card, then hold the card as best you can parallel to your face and at maximum arm length while facing the Moon. Close one eye and use the other to determine the slot into which the Moon fits snugly. Be sure to ink the date and time under that slot.

Tuck the card in a place you'll remember and then bring it out again and compare the Moon's size at apogee. Repeat the procedure and see what slot the Moon fits into *this* time. When conducting your experiment, you'll find it's easier to see the slots if the Moon is viewed in a twilit sky.

Unrelated to the supermoon but inflating its apparent size all the more is the Moon Illusion. Around moonrise and moonset, when the full moon hovers at the horizon, we've all seen that huge orange globe. What makes it so? In daily experience, an object off toward the horizon like an airplane or flight of geese is generally farther away than those directly above us. But for extraterrestrial bodies such as the Moon, Sun and the constellations, which are identical in size whether on the horizon or at the zenith, we have no reference. Therefore, when we gaze at a horizon Moon, which clearly lies beyond every object in the foreground, our brains assume it must be farther away than the overhead version. We compensate for this perception by inflating the Moon's size. In a sense, our brains force the Moon to meet our expectations of how big it should be.

In reality, a rising or setting moon is slightly farther away and a tiny bit smaller than when it's overhead because we have to look across the curvature of the Earth and then out to the Moon. When it's up high, there's no Earth in the way, so we're a little closer to the Moon. The difference amounts to only 1.5 percent, and is overwhelmed by the effect of the moon.

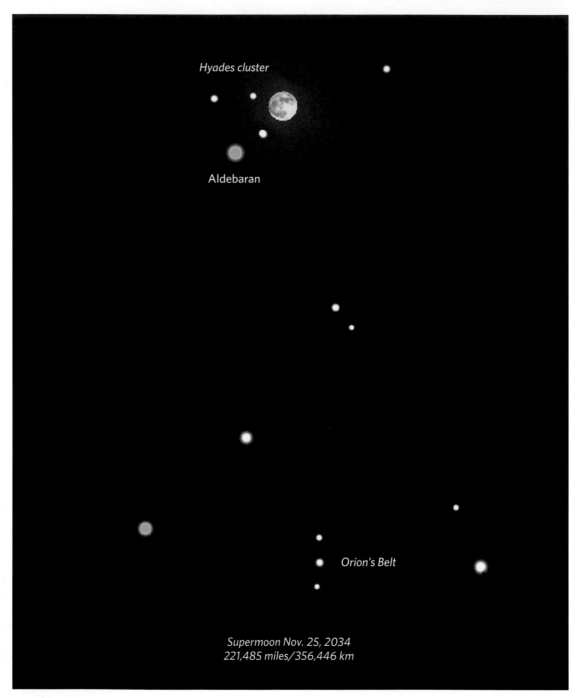

Hyades cluster

Aldebaran

Orion's Belt

Supermoon Nov. 25, 2034
221,485 miles/356,446 km

⌃   *Although supermoons occur every year, some are closer than others. 2016 was the last exceptionally close perigee; the next occurs on November 25, 2034, when the full moon will cross the Hyades star cluster. Source: Stellarium*

How to see a Supermoon

Use the list below to find the date (Eastern time) of the next Supermoon. For the biggest impact, watch either at moonrise or just before moonset. Since more of us are up and around during early evening hours, find a spot with a great view to the east. In winter, the full moon rises in the northern sky; in the spring and fall, in the east; and in the summer, in the southeast. I've included Minimoon (apogee moon) dates so you can compare Moon sizes with your Supermoon sighter.

Supermoons

- 2019—January 21/February 19
- 2020—March 9/April 7
- 2021—April 26/May 25
- 2022—June 14/July 13
- 2023—August 1/August 30
- 2024—September 17/October 17

Minimoons

- 2019—September 13
- 2020—October 30
- 2021—December 18
- 2022—December 7
- 2023—February 5
- 2024—February 24

RESOURCES

- Moonrise/moonset calculator: timeanddate.com/moon/
- Lunar perigee and apogee calculator: fourmilab.ch/earthview/pacalc.html
- Moonrise, a free app for iPhone: itunes.apple.com/us/app/moonrise/id390606204?mt=8
- LunaSolCal Mobile for Android: play.google.com/store/apps/details?id=com.vvse.lunasolcal&hl=en

35

Luna's "Dark" Side

Even if you live in a big city, you'll get your eyeful of earthshine, so long as you can spot the waxing crescent moon at dusk or waning moon at dawn. Sunlight has ways of getting around, and one of them is by reflection. It streams across the solar system, lighting up every planet, moon, asteroid and comet. Objects closer to the Sun not only feel its heat more intensely, all things being equal, they appear brighter, too. We see the Moon and planets because they reflect sunlight, otherwise they would be invisible. Shut down the Sun, and the planets would wink out in the order of their distance from Earth. First the Moon, then the inner planets, and lastly, Neptune.

When the Moon returns to the western sky at dusk, sunlight illuminates just the bottom edge and we see a thin crescent. But even a casual glance will show there's something more going on here—the entire outline of the Moon, though it looks gray and faint. For generations, people have called it "the old moon in the new moon's arms."

The description is a delightfully poetic way to describe **earthlight**—sunlight that reflects off the Earth, out to the Moon and back again to our eyes. Earth's no perfect mirror. Some of the sunlight

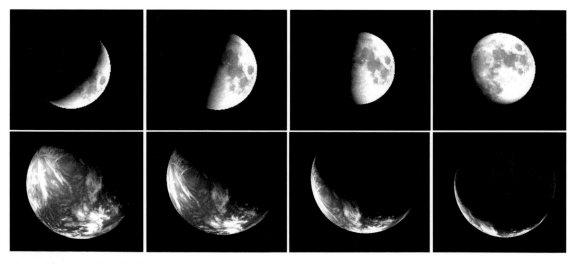

▲ *Moon and Earth phases are complementary. From a crescent moon, the Earth appears almost full. Sunlight reflecting off our globe lights up the dark, shadowed portion of the Moon with earthlight. Credit: Bob King with images from Stellarium (top) and NASA*

is absorbed by water and ground, so only a portion reflects out into space. When it reaches the Moon 1.3 seconds later, the dark lunar soil greedily absorbs much of Earth's reflected light and sends back only a little in return.

After two reflections and a journey of some 480,000 miles (772,500 km), or two Moon distances, only a modest trickle remains, the reason the Earth-lit Moon looks faint and ghostly compared to the Sun-lit crescent. Imagine you're an astronaut standing somewhere on the Earth-lit Moon. Ground illumination would resemble deep twilight here on Earth with the stars shining bright and steady in the black sky overhead. Among them, brighter and bigger than anything else in the heavens, Earth would shine down. No wonder you can see the landscape so clearly. When the Moon's a crescent back on Earth, up at the Moon, an astronaut sees a nearly full Earth hanging in the sky. And just as the full moon lights up the night brighter than any other phase, a full Earth does the same at the Moon. Earth is nearly four times the full moon's apparent diameter and 50 times brighter.

Earth and Moon phases complement each other. When we see a lunar crescent, our alter ego on the Moon sees a gibbous Earth. When the Moon is half, Earth is half, and when the Moon is full, Earth's a skinny crescent. As the crescent moon waxes, Earth wanes in the Moon's sky, and the light reflected from our planet and delivered to the Moon lessens, causing the earthshine to fade. That and the Moon's own increasing brightness as it fills out make it difficult to see earthshine beyond half-moon phase.

▲ *Earthlight morning—Earthshine dimly illuminates much of the waning crescent moon before dawn the same way it does the waxing crescent at dusk. The dusky light is twice-reflected sunlight. Credit: Bob King*

If you make a point to look for it, earthshine is visible up to 5 days past new moon without optical aid. Binoculars or a small telescope extends that to about eight days. My personal best is nine days past new with a 10-inch (250-mm) telescope. When using binoculars or a telescope: Place the glary half of the Moon out of the field of view to let your eyes better see the much weaker earthshine.

When the crescent's thin, take some time to roam the lunar twilight zone with a telescope or binoculars. You'll spy half a dozen or more lunar seas and the bright-rayed craters, Tycho, Copernicus and Aristarchus. Studying the Moon's "dark side" feels like finding your way at night using only starlight. It's an adventure, plus we get a sneak peek at features that won't be in sunlight for days.

While earthshine is visible anytime the Moon's a crescent, northern hemisphere observers will see it best at dusk in the spring months and at dawn in the fall. Then, the crescent stands highest in the sky and is least affected by horizon haze. Because the brightness of earthshine depends primarily on Earth's cloud cover, variations in its intensity may help us understand changes in our planet's climate.

Earth and the Moon share an ancient bond. According to prevailing theory, the Moon originated in a catastrophic collision between a Mars-size planet and the early Earth 4.5 billion years ago. The impactor and a good chunk of the planet were pulverized into a molten hail of rock and dust that later coalesced into the Moon. Seeing the light touch of Earth on the lunar landscape reminds us of the deep connection between the two worlds.

How to see earthshine

Use the Moon calendar link or a mobile phone app (see below) to find out the time of new moon. Crescent moons appear for several evenings and several mornings in a row every month of the year on either side of new moon. As described earlier, the spring and fall are best. Earthshine is visible on the thinnest crescents but much easier to see when the Moon is between two to four days past new (evening) or two to four days before new (morning). Then, the Moon is higher up in a darker sky with improved contrast between the Earth-lit portion and the sky background.

RESOURCES

- Moon phases calendar: moonconnection.com/moon_phases_calendar.phtml
- The Moon—Calendar Phase of Moon. Free app for iPhone: itunes.apple.com/us/app/the-moon-calendar-phase-of-moon-free/id1077455978?mt=8
- Phases of the Moon for Android: play.google.com/store/apps/details?id=com.universetoday.moon.free&hl=en

36

Uranus and Neptune

You may have glimpsed these in binoculars when you took on the challenge of "seeing all eight planets in one night." Now, you'll want to see these cold, remote worlds up close through a telescope. *Close* is a humorous term when it comes to astronomy. Depending on your perspective, "close" could be anywhere from a quarter million miles away (the Moon's distance) to 26 trillion miles, the distance to Alpha Centauri, the closest star system, or 2.5 million light-years, the distance to Andromeda, one of the *closest* galaxies. See what I mean?

Uranus and Neptune lie 1.8 and 2.8 billion miles (2.8 and 4.5 billion km) from the Sun, respectively, and both are nearly four times bigger than the Earth. Astronomers call them "ice giants" in contrast to the gas giants, Jupiter and Saturn. Ice giants are composed primarily of elements heavier than hydrogen and helium and are rich in water, ammonia and methane ices.

Methane is an odorless gas most of us know as "natural gas" and commonly used to heat our homes. Because it's odorless, gas producers add an odorant to help alert us to potential leaks. Cows produce lots of methane through belching and flatulence, but whatever *that* smells like isn't methane's fault.

Uranus   Earth   Neptune

⌃ *Uranus and Neptune possess thick atmospheres and are considerably larger than the Earth but, because of their great distance, appear as minute disks in amateur telescopes. Source: NASA*

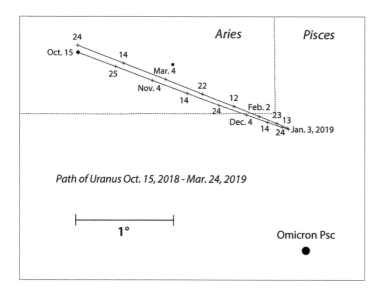

Path of Uranus Oct. 15, 2018 - Mar. 24, 2019

Composed of carbon and hydrogen, methane colors both planets lovely shades of aquamarine blue. The gas condenses in their bitter cold atmospheres as clouds that absorb red light and reflect back blue. When you point a small telescope at either planet and crank up the magnification to 100x to 150x, each becomes a tiny blue dot.

Uranus looks the size of a pinhead and Neptune a little larger than the period at the end of this sentence. If nothing else, seeing them through the eyepiece gives you a real sense of how remote they are. It also gives us new respect for William Herschel, the English astronomer who stumbled across Uranus on March 13, 1781. Because most everyone considered the number of planets settled at six, the idea of a new one wasn't the first thing to cross his mind. Herschel figured it must either be a comet or an unresolved double star. Further observations by him and other astronomers showed the object for what it was: the first new planet discovered since antiquity.

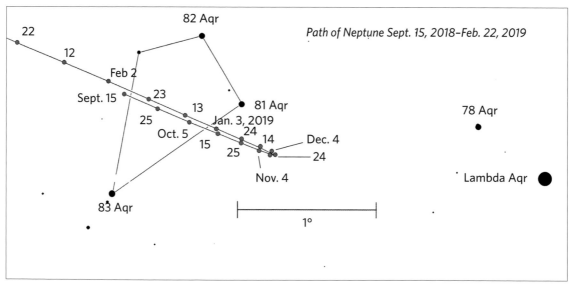

▲ *Use this map to track down Neptune near Lambda Aqr in the 2018–19 season. The planet's position is marked every 10 days and stars are shown to magnitude 8.5. Source: Chris Marriott's SkyMap with additions by the author*

To bolster your hope of finding the planet, Herschel made his discovery using only a 6.2-inch (157-mm) telescope. With a map showing where to look, you should be able to nab it in binoculars.

Uranus soon became a bit of an embarrassment because irregularities in its orbital motion suggested that the gravity of another more distant body was to blame. Two mathematicians, Urbain Le Verrier of France and John Couch Adams of England, took up the task of calculating the position of the hypothetical planet based on the anomalies.

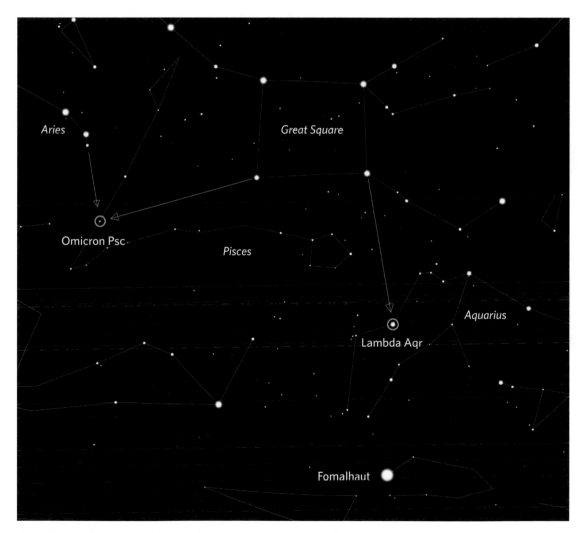

^ *The two key stars for finding the outer planets in fall and winter 2018 are Lambda Aquarii for Neptune and Omicron Piscium for Uranus. Use the Great Square asterism and the bright star Fomalhaut to help you navigate to them. Source: Stellarium*

Working with Le Verrier's calculations, German astronomer Johann Gottfried Galle peered through the telescope at the Berlin Observatory on the night of September 23 to 24, 1846, ticking off stars near the predicted position, hoping to find one that wasn't supposed to be there. To his disbelief, he hit upon it just after midnight, less than 1° from Le Verrier's predicted position. Contemporary physicist François Arago cleverly summed up the remarkable discovery, crediting Le Verrier as the man who "discovered a planet with the point of his pen."

Both blue-eyed planets are visible with binoculars, though a telescope is really the way to go; they'll show disks and color, things you won't see in ordinary binoculars. Because they're far from the Sun, the chilly worlds move slowly across the sky, with Uranus completing an orbit every 84 years and Neptune in 165 years.

Think about that for a moment. Uranus goes around once in about the same time it takes one of us to "go around" this life. Neptune, meanwhile, has only completed a bit more than one orbit since its discovery.

The charts will point you to the planets through the 2018–2019 observing season. You can also make your own finder maps with Stellarium or use a phone app to pinpoint their positions any time in the future.

A tip on the phone apps—remember to zoom in so that you can see all the stars of similar brightness around either planet. Just like Galle, you'll need to eliminate them to be sure you're looking at the planet and not a star. If you're in doubt about a planet's identity when using a telescope, increase the magnification. Stars look like tiny points with small spikes, while planets are clearly defined disks.

How to find Uranus and Neptune

You don't need pristine skies for these guys; the suburbs will do. For a more pleasing view I suggest 40-mm or larger binoculars that will show both Uranus and Neptune on moonless nights. A modest 4.5- or 6-inch (115- or 150-mm) telescope and an eyepiece that will magnify 100x or higher. Through the early 2020s, Neptune will track across eastern Aquarius, which is best visible during evening hours from September through December. Uranus plods across Aries during the same time frame and is well placed for evening viewing from mid-October through February.

RESOURCES
- Stellarium or a phone app like Sky Chart described on page 10.

37

Crescent Venus

At some point in our busy lives, on the way to work, school or waiting in line at the landfill, we were momentarily distracted by the sight of a brilliant "star" in the twilit sky. Almost certainly it was Venus. No other star-like object shines as steady and bright, and none will until the day of that long-overdue supernova. For now, it has no competition.

Because the second planet from the Sun orbits inside Earth's orbit, we periodically get to see it pass between us and the Sun. As you can demonstrate for yourself by taking a ball and holding it between you and a bright light source, no light reaches the side of the ball facing you. But if you move the ball to the left or to the right of the source, some light now shines along the rim of the ball in the shape of a crescent. That's the shape you get when you sidelight a spherical object, and a demonstration of why Venus, just like the Moon, goes through phases.

⋀   *When Venus is in crescent phase, it's easy to see its shape in a pair of binoculars. Credit: Sharin Ahmad*

To visualize the other phases, imagine you and three friends standing in a circle 90° or a quarter-circle apart. A lamp, representing the Sun, shines at the center. You're holding a big, white ball—the planet Venus—directly in front of you that blocks the light from the lamp. As before, you see the unlit side of Venus ("1" in the illustration). If you move the ball slightly to the right, a crescent of light illuminates the left edge. Move it left, and the crescent appears on the right.

Now, toss the ball to the friend on the right. When he holds the ball out toward the lamp (2), from your perspective, its left half is lit by the lamp and looks like a half-moon. That person tosses the ball to the person on their right, who stands on the opposite side of the lamp from you. When he holds the ball toward the lamp (3), you see a "full-moon" Venus. He next tosses the ball to

⋏ *As we toss our "Venus ball" from one person to the next around the lamp, both its phase and apparent size change. At "1," the ball appears as a thin crescent in near-silhouette; at "2," a half-moon; at "3," it's in full phase and at "4," half again. Credit: Gary Meader*

the friend on his right (4). Once again, from your perspective, Venus is half-lit, but this time, it's the right side. Finally, he tosses the ball back to you, and you end as you began, holding a dark, "new-moon" Venus.

Congratulations! You've just put Venus through all its phases. If you paid close attention, you also noticed that when Venus was close to you during its "new" and crescent phases, it looked *larger* compared to the full-moon Venus, which was on the opposite side of the lamp and farther away.

As in our experiment, so in real life. When Venus is a crescent, it appears much bigger than at anywhere else along its orbit. So big, you don't even need a telescope to see its shape. Binoculars magnifying just 7 to 10x will turn the star-like planet into an exquisite little crescent "moon." The other phases are also fun to see, but only the crescent is close enough and large enough to easily discern in binoculars.

For the best views, I recommend looking at the planet in twilight before it gets too dark. That way you avoid Venus's glare, which makes it tougher to see its shape. The evening crescent opens to the left or east exactly like the waxing crescent moon in the evening sky; the morning crescent opens to the right or west like the waning crescent moon. Occasionally, both the crescent moon and crescent Venus will be in conjunction, and we get that rare treat of seeing two crescents at once.

How to see a crescent Venus

You can use the list on the next page to reference the best views of the evening crescent. In case you'd like to look up times yourself, search online for when Venus is in **inferior conjunction** with the Sun, the time when the planet passes between the Sun and the Earth. The month before and month after conjunction are the best times to hunt the Venusian sliver.

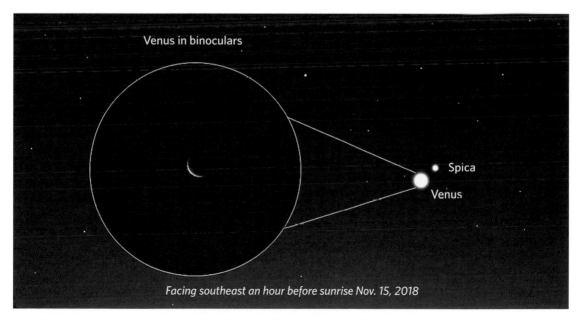

Facing southeast an hour before sunrise Nov. 15, 2018

▲   *You can spy a Venusian crescent at dawn in the southeastern sky in mid-November 2018. On the 15th, it will be close to Virgo's brightest star, Spica. Source: Stellarium*

2018

- Early October **very** low in the southwest after sunset.
- All of November in the east at dawn. Biggest and thinnest in the first half of the month.

2020

- From mid to late May low in the northwestern sky in early evening twilight. Best around the 20th before it sinks too low.
- Mid-June to early July low in the northeast at dawn. Best around June 20, when it starts to climb out of the solar glow.

2021–2022

- Mid to late December 2022 in the southwestern sky at dusk.
- Mid-January to early February 2022 in the southeast at dawn.

RESOURCES

- Curt Renz has a great astro site with a full layout on everything Venus: curtrenz.com/venus.html

- Derek Breit's yearly calendar of astronomy events is also a great place to drop by. You can search there for inferior conjunction dates: poyntsource.com/New/Diary.htm

- Stellarium or a phone app like Sky Chart described on page 10 will show you Venus and provide data on its size and what percentage of the planet is illuminated by the Sun (20 percent or less is best).

38

Cygnus Star Cloud

When I die, please box up my soul and ship it to the Cygnus Star Cloud. Few places in the Milky Way look so inviting on a summer or fall night than this magnificent "island" of stars that spans a fist and a half from the center of the Northern Cross (Cygnus the Swan) to Albireo at its foot.

Even on full moon nights, I can still make out the cloud's faint presence. That's how bright it is. On dark nights away from the city, look closely at this fat oval glow with the naked eye. It's granulated, not smooth and milky. Scores of stars at and near the limit of visibility seem to twinkle on and off along its length like distant fireflies. Hands down, this is the most visually rewarding star cloud visible with the naked eye from the northern hemisphere.

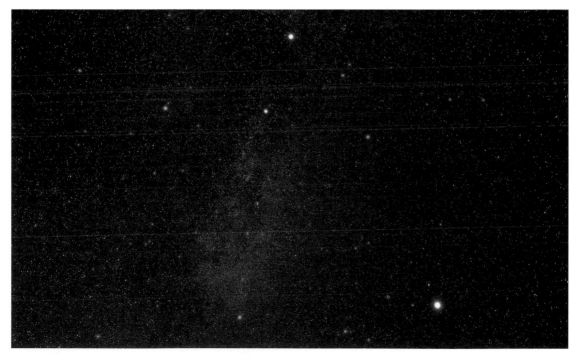

▲ The star-rich Cygnus Star Cloud is one of the brightest sections of the Milky Way and covers two-thirds of the Northern Cross asterism. Binoculars and telescopes show tens of thousands of stars here. Credit: Bob King

Gazing into Cygnus, we peer thousands of light-years down the length of our local spiral arm to where it wraps around the center of the galaxy. Stars near and far overlap to create a region of stellar saturation. And all this from your own backyard. Amazing.

Billions of stars aren't the only reason the cloud stands out so boldly. There's also very little dust down this road. Dust absorbs starlight, the key reason we can't peer into the center of the Milky Way Galaxy with ordinary telescopes.

There's much to glean in the Cygnus Cloud with the naked eye, but don't stop there. Go the full distance with binoculars and telescope. You'll see s-o-o-o many stars with binoculars—way too many to count. My favorite way to explore the region is with the telescope at low magnification. I start near the center of the Cross and sweep west from one side to the other, then lower the scope a degree and sweep again in the opposite direction. Back and forth, back and forth like mowing the grass. Try it yourself.

You'll run into the most beautiful things on every sweep: colored stars, star chains, dark nebulae (apparent vacant spaces between starry billows that in truth are thick clouds of interstellar dust) and several small star clusters. A half hour spent in the Cross's starry sweet shop calms the mind and recharges our sense of wonder.

Other star clouds dot the Milky Way band including a prominent round "puff" in the constellation Scutum the Shield below the Summer Triangle and two bright chunks in Sagittarius. In these we look out toward more remote regions, where the stars are generally fainter and don't stand out as brightly as those in Cygnus.

How to see the Cygnus Star Cloud

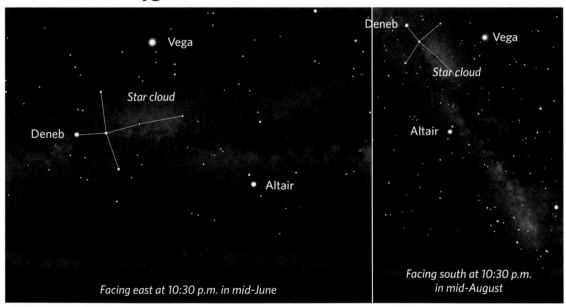

▲   *In spring, the constellation Cygnus the Swan comes up on its side in the eastern sky. By August and September, it stands nearly overhead. Source: Stellarium*

Choose your equipment—naked eye, binoculars, telescope or all of the above—and pick a moonless night anytime from June through November when the Northern Cross is well placed for viewing. In early summer, it lies on its side in the eastern sky with the band of the Milky Way horizontal to the horizon. By August, Cygnus rides high in the southern sky near the zenith with the Milky Way diagonal to the horizon. Come fall, the Cross stands straight up and down in the northwest, making a "last stand" before knuckling down for winter.

Any binoculars will do, but widefield types show a bigger piece of sky and more stars at a glance. A 50-mm glass gathers more light than either a 25-mm or 35-mm pair and will show more stars. I recommend 10x50 wide field binoculars. Any telescope will show even more but consider a 6-inch (150-mm) for brighter views of the nebulae and star clusters you'll run across.

RESOURCES

- Moon phases/lunar calendar. Phase times are shown for your time zone and account for Daylight Saving Time. Very handy! timeanddate.com/moon/phases/

Magellanic Clouds

Small asteroids orbit big asteroids; moons orbit planets; planets orbit suns; suns orbit other suns; small galaxies orbit bigger galaxies and small galaxy clusters orbit bigger galaxy clusters. Everything really does go 'round and 'round.

The Milky Way is one of more than 54 galaxies in the Local Group, a cluster of mostly small galaxies held together by the gravity of its members. Andromeda and the Milky Way are the group's heavyweights; at least two dozen smaller satellite galaxies orbit the Milky Way the way the planets orbit the Sun.

Only two of them are bright enough to see with the naked eye and both are visible from the southern hemisphere: the Large and Small Magellanic Clouds. They look like pieces of the Milky Way that split off and drifted away to other parts of the sky. With binoculars and telescopes, we can mine an assortment of celestial treasures in each.

∧   *From the southern hemisphere, the Large and Small Magellanic Clouds resemble detached portions of the Milky Way. The bar in the larger cloud (left of center) is obvious in a dark sky. Credit: Joseph Brimacombe*

They're named for Portuguese explorer Ferdinand Magellan, who first reported the sight during his circumnavigation of the globe from 1519 to 1522. In January 1521, in *The First Voyage Round the World, by Magellan*, Magellan's assistant Antonio Pigafetta wrote: "On the South Pole can't be seen the same constellations as on the North Pole. Here there are two groups of small misty stars, which resemble little clouds, which are not much remote from each other."

These "misty stars" have always been part of the sky lore of Australian Aboriginal peoples, the Polynesians and other cultures. Sitting around a campfire in the Outback a thousand years ago in Western Australia, you would have heard the story of how the Clouds represented the camp of an older couple too ill to gather their own food. The Large Cloud symbolized the man; the Small Cloud, the woman. A bright star near both, perhaps Canopus, was their fire.

Both galaxies adorn the southern hemisphere evening sky from about mid-September through mid-April though they're highest and easiest to see in that hemisphere's midwinter. While they can be glimpsed low in the southern sky from 10° north latitude (Caracas, Venezuela), you'll get a much better view if you pick a spot well below the equator. Argentina, Zimbabwe or Australia are ideal locations.

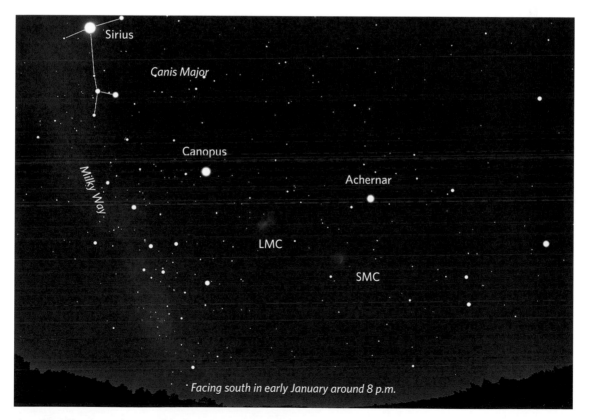

Facing south in early January around 8 p.m.

▲ *The Clouds are best visible from January through early April from the Southern hemisphere. A line from Sirius, the brightest star, through Canopus will take you to the LMC and from there to the SMC. Source: Stellarium*

The Large Magellanic Cloud (LMC) is about 160,000 light-years away and straddles two constellations, Dorado the Goldfish and Mensa the Table Mountain. Its neighbor, the Small Magellanic Cloud (SMC), lies about 200,000 light-years from Earth in Tucana the Toucan. Both are classified as irregular dwarf galaxies and orbit the Milky Way once about every four billion years. The gravity of the far more massive Milky Way has raised tides in both galaxies, like the Moon raises tides on Earth, and distorted their shapes. In a tit for tat, the Clouds have yanked on and likewise distorted the outer parts of our galaxy's disk. Cold streams of hydrogen gas connect all three, pulled from each other by their mutual attraction.

Both Clouds lie about 20° or two fists from the south celestial polestar, Sigma Octantis, so they circle about it and never set for observers living south of about 20° south latitude. From a dark sky, the LMC looks like a fuzzy patch of light, brighter toward the center, and spans some 7°. One balled fist held at arm's length will cover it with room to spare. The SMC is less than half as large, about 3° across, and two fists (20°) to the west of its bigger sister.

Seeing both with the naked eye is a fine accomplishment in itself, because each is one of the most remote objects visible without optical aid. But binoculars and especially a small telescope will show dozens of star clusters and nebulae with the greater concentration of riches in the LMC. The LMC is also home to the Tarantula Nebula, the most active star-forming region in the entire Local Group of galaxies. Easily visible to the naked eye as a brighter spot within the galaxy, binoculars show a north–south extended glow, which expands into a mass of loops and whirls with a modest-size telescope.

Amateur astronomers living in the southern hemisphere consider the Tarantula one of the most exquisite nebula in the sky. To give you an idea of its size, if it were located at the same distance as the Orion Nebula (1,350 light-years), the Tarantula would cover an area of sky twice as large as the Big Dipper. The dense cluster of newborn suns at its core would shine bright enough to see in

⌃ *The photo, taken with a telephoto lens, shows the LMC's central bar of stars and the pink glow of the giant Tarantula Nebula. Credit: Gonzalo Vicino/CC BY-SA 3.0*

the daytime. The closest and brightest supernova observed since the invention of the telescope, SN 1987A, exploded into view near the Tarantula Nebula in February 1987 and was easily visible to the naked eye at magnitude 2.9.

I saw the LMC for the first time in the desert near Arequipa, Peru, during a trip to see Halley's Comet in 1986. It looked like part of the Milky Way but off by itself. I just looked up and there it was. Because my friend and I were preoccupied with a broken-down car at the time, I wasn't able to get my telescope on it, one of the reasons both it and the SMC are still on *my* bucket list!

How to see the Magellanic Clouds

Use the Clouds as the seed to plan a trip to Australia. You know you've always wanted to visit Down Under and have dinner in the real Outback, so put the Clouds on your agenda and book a ticket. Australia's high-travel season runs from about December through February, also the best time to see the Clouds. If fares are too high for your taste, consider the October–November "shoulder season." For a good look, you'll need to get outside of big cities like Sydney or Brisbane and into the countryside.

One of the continent's most famous sites, the monolithic Ayers Rock in the heart of the Northern Territory, would make an ideal viewing spot. From there, you can catch both galaxies along with mouth-drooling views of Orion, the southern Milky Way and Southern Cross.

In October, you'll see this amazing sight in the early morning hours before dawn or in January around 10 to 11 p.m. local time. I encourage you to pack good binoculars, both to study the Magellanics and explore parts of the Milky Way you can't see from home. A telescope is even better, but you'll need a good map so you can track down the many deep-sky objects inside each galaxy. I've included a reference atlas in the Resources section.

RESOURCES

- A Visual Atlas of the Magellanic Clouds: webbdeepsky.com/publications/books/visual-atlas-magellanic-clouds. Contact: donjmiles@googlemail.com

- *Sky & Telescope's* Pocket Sky Atlas. Includes general maps of the Clouds and a close-up map of the LMC: amazon.com

- Pair of binoculars, preferably in the 40 to 50 mm range magnifying 7 to 10x

- 6-inch (150-mm) or larger reflecting telescope

40

Magical Mira

▲ *This Hubble Space Telescope image of Mira reveals its odd football-like shape, which may be linked to dramatic changes in the star during its expansion-contraction cycles. Source: Margarita Karovska/NASA/ESA*

Now you see it, now you don't. Every 332 days, the red giant star Mira, in the constellation Cetus the Whale, pulls itself out of a hat and shines in plain view like any other stars in the sky. Then, without warning, it fades and disappears.

Or so it seems. In binoculars and small telescopes, Mira remains faintly visible around, and well below the naked eye limit. About three months later, it rebounds again and returns to its former brilliance. What's going on here?

Mira's an enormous red giant star desperately close to the end of its life. It wasn't always this way. The star began life much like the Sun, happily burning hydrogen in its core and settling in for a long, stable adulthood of steady light. Stability's a good thing because at least in our solar system, it provided the time needed for life to evolve and become both ubiquitous and tenacious on our little blue planet. Did any possible planets around Mira take advantage of the same?

Because Mira was born several billion years before the Sun, it's further along its evolutionary track and provides a glimpse into the Sun's distant future. All of the hydrogen in its core has been cooked into the heavier element helium. Gravity compresses the helium further, increasing the core's temperature until it can burn helium into carbon and oxygen. Simple elements like hydrogen, helium and a bit of lithium were cooked up in the early universe not long after the Big Bang. More complex ones, especially carbon, so important to life, were synthesized in the cores of stars.

Helium burns hotter than hydrogen. As Mira began to burn its helium, the heat produced triggered hydrogen burning in a shell around the core, causing the star to balloon into a red giant. Mira may have started with a similar amount of material as the Sun, but it's evolved into a giant gasbag more than 350 times its size. The wispy outer layers of such enormous stars don't feel much gravitational tug from their distant centers, so they're constantly losing material to space like the perpetual, private dust storm surrounding the Peanuts character, Pig Pen. The huge outflow of material also distorts the star's shape, making it out-of-round. Mira began to expand and contract like a balloon inflated and deflated at regular intervals. Internal changes in Mira, as it adjusted to burning inside and outside the core, made it unstable. Over a period of 332 days, the star's outer envelope expands and cools. Gravity then draws the material back in and reheats it, initiating another cycle of expansion. Mira's brightest when it's contracting, faintest when expanding. Its pulses remind us of the human heart, but instead of 72 beats a minute, Mira beats once every eleven months.

Several thousand Mira-type stars are known, with pulsation periods between 100 and 1,000 days, but Mira was the first discovered. German pastor and astronomy dabbler David Fabricius noticed Mira's variations in August 1596 while using it as a comparison star to determine the position of Jupiter in nearby Aries the Ram. From August 3 to 21, the star's brightness increased. Then, in September, it began to fade and was invisible by October.

A sudden brightening followed by an eventual disappearance is typical of a nova or "new star," exactly what Fabricius assumed he was looking at. But he was taken by surprise when it reappeared again in February 1609. He must have realized something exceptional happened, but for whatever reason, Mira was forgotten or ignored until astronomer Johannes Phocylides Holwarda rediscovered it 29 years later. In 1638, Holwarda determined that the star took eleven months to rise from minimum brightness to maximum and back to minimum.

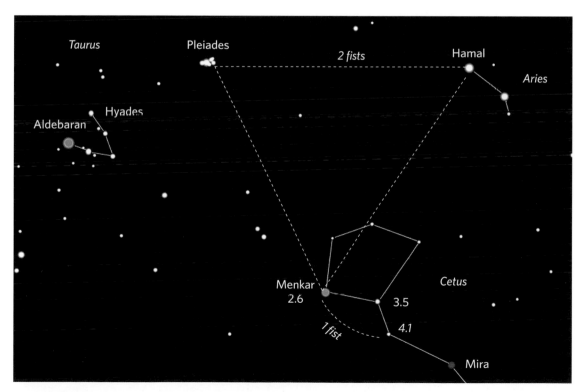

∧　To find Mira, start with the Pleiades and form a triangle with Hamal in Aries and Menkar in Cetus. Mira's about a fist below and right (southwest) of Menkar. Don't be surprised if the star's absent from view at times! Source: Stellarium

By this time, other astronomers were paying attention to the star, including Johannes Hevelius, who in 1642 named it Mira, Latin for "wonderful." Both the star's name and the word "miracle" share the same root.

Pity that Pastor Fabricius never lived long enough for the astronomical community to appreciate his discovery. He died in 1617 at the age of 53 after alleging—from the pulpit—that a local peasant stole one of his geese. Not long after, the man struck him in the head with a shovel and killed him.

The fates will be kinder to you when you step outside to see what Mira's up to. I've included a short list below of the approximate times when the star reaches peak brightness through 2024. For several months around that time, the star is easily visible without optical aid. For the rest of its cycle, a pair of binoculars or small scope will keep it in view. It takes the star about 100 days to climb from minimum to maximum, and about 230 days to fall back to minimum. That means some exciting observing in the weeks before peak, when Mira's brightness changes rapidly.

At maximum, Mira can shine as brightly as the brightest Big Dipper stars, but it's usually closer to magnitude 3 to 3.5 with an orange-red color that's obvious in binoculars or a telescope. Fall and winter evenings are the best times to see the star, but if Mira's at the top of its cycle in midsummer, it's worth rising early to see it before dawn.

How to see Mira

The map will help you navigate to Mira. Although Cetus is rather faint, bright clusters in Taurus will point you in the right direction. The V-shaped Hyades star cluster, located just above Orion's shoulder, points directly at Menkar, the brightest star in Cetus, two fists away. Mira lies 12° or a little more than a fist in the same direction. If you don't see it, that means it hasn't reached naked-eye brightness yet for your location. Mira is normally visible without optical aid about four weeks *before* peak light and up to six weeks *after*.

Approximate dates of peak brightness and best times to view:

- December 16, 2018 (evening sky)
- November 13, 2019 (evening sky)
- October 11, 2020 (late evening sky)
- September 8, 2023 (midnight and early morning)
- August 6, 2024 (early morning)

RESOURCES

- Stellarium or a phone app like Sky Chart described on page 10.
- More on Mira: aavso.org/vsots_mira
- "Will Earth Survive When the Sun Becomes a Red Giant?" from Universe Today: universetoday.com/12648/will-earth-survive-when-the-sun-becomes-a-red-giant/

41

Bright Comet

If anything represents the value of science in shifting humanity's viewpoint from fear and ignorance to knowledge and wonder, it's our understanding of the nature of comets. For centuries, they were considered omens of bad times, whether in the present or soon to come. The last thing you'd want to see after a hard day in the fields was a big, bright comet hanging over your house.

But then came Galileo, Kepler and Newton, and by 1687, the date Newton's theory of gravity was published in his book, *Principia*, comets were tamed. Edmond Halley, a friend of Newton's, paid for the publication out of his own pocket because he realized the importance of Newton's ideas. Then, using Newton's law, Halley calculated the orbits of what were thought to be three separate, isolated comets, only to discover that they were repeat visits of just a single one. In honor of his discovery, it was named Halley's Comet.

Before Halley's time, no one knew comets orbited the Sun like the planets. They'd just appear out of nowhere—like they still do—and scare the heck out of people. You and I are the benefactors of the hard work of Newton and friends because we know that comets aren't evil omens but small, asteroid-like bodies made of ice and dust that orbit the Sun in a manner similar to planets.

▲ *Comet Hale-Bopp was one of the brightest comets in recent decades and easily visible with the naked eye for weeks in the spring of 1997. Traveling on a long, narrow orbit, it won't return to Earth's vicinity again until around 4385. Credit: Bob King*

That leaves us free to enjoy a comet for what it is, a thing of nature transformed by the heat of the Sun into one of the most alluring spectacles we might hope to see in a lifetime. When far from the Sun, most comets are inert orbs of dusty, dirty ice anywhere from a few hundred feet to about 6 miles (9.6 km) across. Close-up photos from spacecraft flybys have revealed that some comets are shaped like bowling pins with a bulbous head and bottom separated by a narrow waist. They orbit the Sun with periods ranging from a few years up to *millions* of years. Some of you may remember Comet Hale–Bopp from 1997, the last exceptionally bright comet to grace the Northern Hemisphere's skies. It's not expected to return for more than 2,500 years.

The reason some comets take so long to go around the Sun is because they move in elongated, cigar-shaped orbits that take them out beyond Neptune and even as far as the Oort Cloud, a huge repository of comets that's been cooling its heels since the formation of the solar system. The Oort Cloud occupies a roughly spherical volume of space centered on the Sun 465 billion to 9.3 trillion miles (748 billion to 15 trillion km) away. If we could shrink the solar system down so the cloud were half a football field (155 ft/47 m) away, the Sun and eight planets would fit in a sphere only a foot (0.3 m) across at its center. At the far end of an orbit, steeped in the bone-cold chill of space, a comet is practically indistinguishable from an asteroid. But let time pass. As it draws closer to the Sun, heat vaporizes water and other dust-laden ices in the comet. Set free by vaporization, the dust gathers in a cloud around the nucleus to form a tenuous atmosphere called the **coma**, from the Latin "hair."

The coma gives a comet its familiar fuzzy appearance, but the real magic happens when the physical pressure of sunlight *pushes* the boiled-off dust back from the coma to form a tail. The dust particles are about the same size as those in the smoke rising from a burning cigarette. It doesn't sound like much to work with and mass-wise it isn't, but when a comet passes near the Earth and we get a sidelong view of the tail, even folks who don't look at the sky much will go outside for a look. Tails give comets their special beauty.

Sunlight illuminates the dust in a comet's tail the same way it lights up dust in the bedroom when you fluff the sheets. Often, a second tail called the **ion tail**, forms when ultraviolet light in sunlight rips away electrons from the atoms of gases in the coma, turning them into ions.

Ions get swept up by magnetic forces embedded in the solar wind, a stream of ionized particles streaming daily from the Sun, and carried away from the coma to form a narrow tail directly opposite the Sun. The most commonly ionized gas is carbon monoxide, which scatters blue light, so ion tails typically look pale blue compared to the yellow-hued dust tails. Changing viewing angles as both Earth and comet orbit the Sun can shift the tail this way and that. Fluctuations in the solar wind can snap off an ion tail and regrow it days later. Watching these changes makes for compelling viewing, the reason some amateur and professional astronomers specialize in comet watching and study nothing else.

Astronomers have plotted the orbits of nearly 200 *periodic* comets, the ones that circle the Sun in fewer than 200 years and have been observed during more than one perihelion passage. A comet or planet is at perihelion when closest to the Sun. Halley's Comet, with a period of 76 years, is a good example of a periodic comet. There are hundreds of nonperiodic comets as well, many of which have only been observed once because they take thousands of years to orbit the Sun.

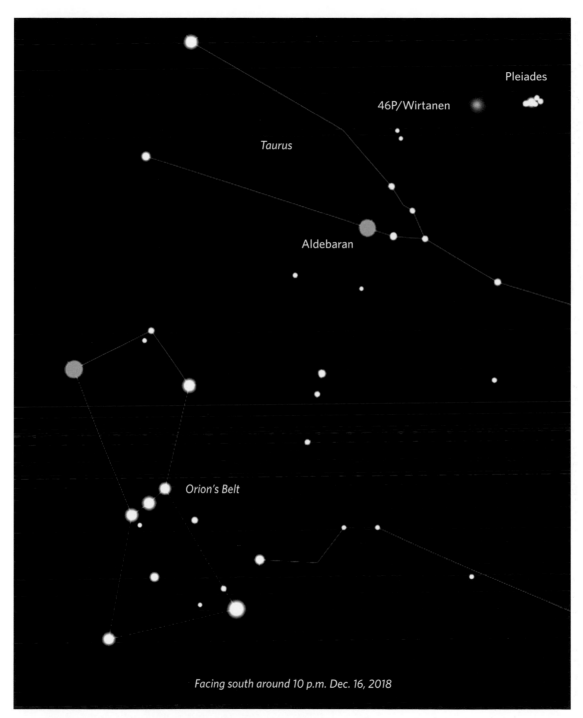

Pleiades

46P/Wirtanen

Taurus

Aldebaran

Orion's Belt

Facing south around 10 p.m. Dec. 16, 2018

▲ *Look for comet 46P/Wirtanen to reach naked eye brightness in December 2018, when it will be near the Pleiades star cluster. Source: Stellarium*

Professional robotic sky surveys with the assistance of students and professors pick off most of the new comets that come winging by every year, but dedicated amateurs occasionally beat the pros and find a few, too. Most are faint, some are visible in amateur telescopes and others bright enough to see with binoculars or, rarely, with the naked eye.

That brings us to the question—how exactly do you *plan* to see a bright comet? It's a tricky business for sure. For all I know, a new comet will be discovered as you read this and pass spectacularly close to Earth after swinging near the Sun, exactly what happened with Comet Hale-Bopp in 1997 for Northern Hemisphere observers and with Comet McNaught in 2007 for Southern Hemisphere viewers. The latter became brighter than Venus.

I've observed nearly 400 comets, most of which have required a telescope. Many were strikingly beautiful, but only a handful could truly be called **"bright"** or **"great"** comets like Comets Bennett (1970), West (1976), Halley (1986), Hyakutake (1996) and Hale-Bopp (1997).

Except for Halley, all of these were drop-ins or unexpected visitors from out beyond Neptune making their first appearance in modern times. No one could have predicted their arrival. First and foremost we need to be patient. If you want to see a bright comet, stay in touch with online comet news, and then make the effort to be out when the comet's out, whether that's in the evening or during morning twilight or the middle of the night.

Any comet, including the periodic ones, might experience a sudden "outburst" and brighten by many magnitudes, but no one can predict exactly when. That said, there are a couple of periodic comets that are expected to put in nice appearances in the years ahead. One is Comet 46P/Wirtanen ("46" means the 46th comet to have its orbit determined and "P" stands for periodic). On December 16, 2018, 46P will pass just 7.2 million miles (11.6 million km) from Earth. It should become easily visible with the naked eye, shining around magnitude 4 near the Pleiades star cluster. No other periodic comet is expected to break the naked eye limit through 2022.

Looking deeper into the crystal ball of time, Halley's Comet (1P/Halley) makes its next perihelion passage on July 28, 2061, when expected to become a little brighter than at its previous return in 1985–86. It may resemble a Big Dipper star with a nice, long tail low in the eastern sky at dawn in late June that year. Come early July, my then very-old children will see it neatly paired up with Venus at dusk.

How to see a bright comet

When the time is right, the map will guide you to 46P/Wirtanen, well placed for viewing in the 2018 winter evening sky, when it should be visible with the naked eye. Use the links below to keep up on comet news, so you can spring into action when a bright one makes its appearance.

RESOURCES

- Visual Comets of the Future (northern hemisphere): aerith.net/comet/future-n.html
- Comet mailing list/latest comet observations and news by dedicated amateurs. Subscribe at: comets-ml-subscribe@yahoogroups.com

42

Airglow

Airglow is a faint, streaky or banded glow visible in the lower half of the night sky from a truly dark location. Never heard of it? I can help change that. Natural airglow from our planet's atmosphere informs us about yet another relationship our planet has to the Sun, but one we only see at night.

Even on the darkest nights from the middle of the Australian Outback or in the hinterlands of Nebraska, you can still find your way at night. Assuming you've been out long enough for your eyes to dark-adapt, you can look around and see all sorts of stuff—your hands, trees, cars and even daisies, when they're in season.

That light has to come from somewhere, and if you guessed it was the stars, you're about 50 percent correct. Stars contribute about half of the night's light. The rest comes from glowing air high overhead called **airglow**. During the day, energetic ultraviolet light from the Sun breaks apart molecules of oxygen and nitrogen—or pries electrons from their atoms—in the upper atmosphere some 60 miles (96 km) overhead. At night, the atoms recombine and regather their electrons, releasing energy as tiny spurts of red and green light as they return to their previous "relaxed" state.

I remember my first encounters with airglow. On otherwise cloudless nights from a dark-sky site, I'd see broad streaks of what looked like cirrus clouds. Yet nighttime weather images showed absolutely no clouds. Later, I learned that airglow could be seen with the naked eye, and I confirmed my suspect clouds by taking time exposures at high ISO settings with a digital camera. Sure as sunshine, green patches and waves glowed on the back display.

Astronauts get the best view of all. As they peer out the windows of the International Space Station, airglow forms a thin, green band that envelops the Earth like the amniotic sac around a developing fetus. In a wonderful natural coincidence, the color of light emitted by excited oxygen is the same as the plants that produce it in the first place—green.

Here's another crazy thing about airglow. Even though it's similar to the aurora in that atmospheric atoms get pumped up by the Sun then dump that energy as light, airglow isn't restricted to the polar regions but visible anywhere on Earth. Equator? Yep. Rural Oklahoma? Yep. Chicago? Nope, too light-polluted.

Airglow varies by time of night, season and solar cycle but appears brightest (if you can call it that!) 15 to 30° or about one to three fists above the horizon, where our line of sight passes through more air compared to when we look up overhead. The more air, the more glow . . . to a point. Haze present nearer the horizon snuffs it out.

Airglow is brightest around the time of peak solar activity called **solar maximum**, which occurs every 11 years with the next maximum expected about 2024. Cosmic rays, which are high-speed, subatomic particles (mostly protons) racing in from all corners of the galaxy, make a contribution to airglow, too. They travel at a significant fraction of the speed of light and strike and excite atoms and molecules in the upper atmosphere.

Another aspect of airglow barely visible to the eye but astonishing in time exposures is the lasagna-like layering of the light set into motion by gravity waves. Wind flowing over mountain ranges, jet stream shear and even powerful thunderstorm updrafts can create atmospheric waves that reach high altitudes and turn the airglow into repeating swells of light. It looks truly eerie.

If you had magical powers and could remove the stars, including the band of the Milky Way and the zodiacal light, the nights would be darker, but airglow would still provide enough illumination to see your hand in front of your face. Even on the darkest nights, we can't escape the Sun's touch.

How to see airglow

Naturally, you'll need a dark sky, or at least half of your night sky should be free of light pollution. If the summer Milky Way looks puffy and textured or you can see the winter Milky Way with ease, you should search for airglow.

▲   *"Rivers" of green airglow appear to flow from the band of the Milky Way on a dark July night. Airglow is visible from any moonless rural location on many nights of the year. Credit: Bob King*

Give your eyes a good half hour to fully adapt to darkness. An hour's even better. I like to spend the time on other projects like looking at star clusters and double stars with my telescope, taking pictures or sweeping favorite parts of the sky with binoculars. When you're ready, look up and scan the sky top to bottom, left to right and right to left, especially in the zone about two to three fists above the horizon. You can scan higher, too—I've detected airglow up to 50° altitude (five fists).

Do you have a tripod and digital camera that can take time exposures? With it, you can verify your observation with a photograph. Open the lens as wide as it can go to let in the most light, usually f/2.8 to f/4.5. Then, select a high ISO from 1600 to 6400. Don't worry about quality. You're using the camera only as a research instrument.

If you don't see airglow the first time, try again another dark night. My evidence is only anecdotal, but the phenomenon seems more obvious in the summer than winter. What do you see?

RESOURCES

- The New World Atlas of Artificial Sky Brightness: cires.colorado.edu/Artificial-light

- Moon phases calendar: moonconnection.com/moon_phases_calendar.phtml

- International Space Station images. Flip through these for some cool night photos that show cities, auroras and airglow: nasa.gov/mission_pages/station/images/index.html

43

Earth-Grazing Meteor

It began innocently enough. I was taking photos of the Orion Nebula on a November night a couple years back when I spotted a meteor low in the northeastern sky out of the corner of my eye. It rose straight up into the sky, its head glowing the same orange color as a lit cigarette. Normally, a meteor burns out in a second or two, barely enough time to alert the person next to you to look up. This one kept going: one second, two seconds, three seconds, four seconds . . . all the while rising higher and higher until it crossed the meridian headed southwest. It finally fizzled out low in the southwestern sky.

I don't recall exactly how many seconds it burned—maybe fifteen—but it was enough time to repoint my camera to the anticipated path of this turtle-speed shooting star and take a picture. While not the brightest meteor I'd ever seen, its duration made it as impressive as any fireball. As if reluctant to say "good-bye," the visitor left a long-lasting trail of glowing air in its wake that lingered for many more seconds after its disappearance.

What I saw was an **Earth-grazer**, a meteor that skims the Earth's atmosphere, skittering across it like a stone skipped on a pond. Instead of burning up in a sudden death flash, this crumb of comet dust slowly scraped itself into glowing oblivion. Some Earth-grazers don't burn up completely but merely graze the air and then return to space. For all I know, this was the fate of my meteor. Is it still up there orbiting the Sun, prepping for Act II?

Tracing its path backward, it was clear my meteor belonged to the Leonid meteor shower, which was peaking that night. The radiant or point in the sky from which the Leonids radiated just *below* the northeastern horizon, the ideal place for an Earth-grazer to make an appearance. Shower meteors strike the Earth's atmosphere *horizontally* at that time instead of raining down from higher up. Because an incoming meteoroid—the solid object that causes the streak—only "kisses" the air throughout its lengthy path, it can keep on exciting air molecules to glow until it either breaks apart, burns up completely, or finally exits the atmosphere.

In case you were curious, a meteor or "shooting star" isn't a flame so much as a tube of glowing air several feet (1 m) across and several miles (5 km) long. Friction from a meteoroid slamming into the atmosphere at tens of thousands of miles per hour causes it to heat up and glow. The particle also compresses and briefly heats the air to more than 3,000°F (1,650°C). Heating springs electrons loose from the trillions of air molecules; when they reattach, energy is released as a flash of light we see as the meteor streak. Sometimes meteors leave a fainter glowing trail called a **train** that can last from several seconds to several minutes, also caused by excited air molecules.

During a shower, multiple meteors per hour are visible depending on the shower's strength and your skies. For the best meteor watching, I always advise catching the peak of a meteor shower, when the radiant is as high in the sky as possible and the meteors most plentiful. Most often, that happens after midnight, making it a challenge for some would-be meteor watchers. Lots of working people and students find it difficult to get up at 3 a.m. for a couple hours, fall back asleep and then start work or school a few hours later. Kids' early bedtimes also can be a problem.

And therein lies the secret solace of Earth-grazer gazing. You can start watching during the early evening when the radiant is either low or just below the horizon. Because every star and constellation rises in the east, face that direction at nightfall when an expected meteor shower is at or near maximum and cross your fingers. You might spot a meteor that just won't quit.

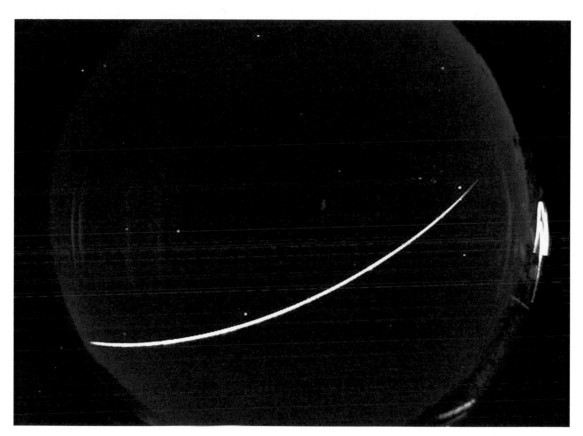

▲   *This spectacular Earth-grazing fireball appeared in the sky over Columbia, South Carolina, on May 15, 2014. After skimming the upper atmosphere for some 290 miles—an exceptionally long distance for a meteor—it finally burned up over Tennessee. It was recorded by one of NASA's all-sky cameras. Source: Marshall Space Flight Center/Bill Cooke*

How to see Earth-grazers

Look up one of the bright meteor showers on page 82 and plan your observing during the evening hours. Some showers like the Eta Aquariids of early May peak just before the start of morning twilight. The radiant is low in the southeastern sky for northern hemisphere observers at that time, the reason this shower is known for Earth-grazers. But you'll have to get up at the odd hour to see it. Most other major showers—the Lyrids, Perseids, Orionids and Geminids—have the potential to spitball an Earth-grazer *before* the clock strikes midnight.

RESOURCES

- Meteor shower list: amsmeteors.org/meteor-showers/

- MeteorActive app for iPhone (free). Great information about upcoming showers along with Moon phases and more expected at the time of the shower. For iPhone only: itunes.apple.com/us/app/meteoractive/id1205712190?ls=1&mt=8

- *Night Sky with the Naked Eye* by Bob King. More information about meteors and meteor showers

Sparkling Sirius

▲ *All stars twinkle, but Sirius, the sky's brightest nighttime star, is the twinkliest. On nights of turbulent air, it puts on an electrifying show. Credit: Bob King*

Our descendants living in colonies on the Moon will look up at the sky and never see a star twinkle. Every object in the heavens will shine with the steadiness of a tiny LED light. Just the thought of it makes amateur and professional astronomers giddy. For us, twinkling is a nemesis, a star's way of broadcasting atmospheric turbulence. Turbulence smears and blurs the delicate details we seek to eke out when observing a planet or nebula through the telescope. Observing from the Moon, where there's virtually no atmosphere, means seeing everything as sharp and still as a painting.

My wife routinely asks me why, after having looked at Jupiter or the Orion Nebula 388 times, I need to see it again. What's to be gained? While there are many reasons to return to the beauties of the night sky, one of them is the hope for serene and steady air, what astronomers call "good seeing." On those few nights a year when it's near perfect, everything comes alive in the telescope and looks utterly real. No exaggeration, it feels like being there.

But I'm not so hardened that I grimace every time a star flutters. I like twinkling not only for what it tells us about the state of the atmosphere but also because it's beautiful to see. The stars are like little firecrackers up there, and the best firecracker of them all is Sirius the Dog Star in the winter constellation, Canis Major the Greater Dog.

All stars twinkle, but our eyes can only detect the shimmering on the brightest suns down to about magnitude 2 or 3 or a little fainter than the stars in the Big Dipper. Sirius is exceptional because it's not only the brightest star in the heavens but also more affected by the atmosphere—at least for observers living in midnorthern latitudes—because it never climbs very high in the sky.

Twinkling, also called **scintillation**, is caused by the churning of small air cells of different density and temperature between you and the star. Warm air is less dense than cold air. Because air temperature decreases 3.5°F for every 1,000 feet in altitude (6.4°C/km), nature guarantees a constant and generous supply of hot and cold air cells.

Each air cell acts like a lens, randomly refracting a star's light this way and that, altering both the star's brightness and position. This can happen quickly as winds blow fresh pockets of variegated air across our field of vision. Although it can feel perfectly calm at ground level, the turmoil above can make a star fluctuate wildly.

When stars are high in the sky, they twinkle less because there's a lot less air in the way. Our gaze passes through 10 miles (16 km) of thick air in the lower atmosphere and then literally into thin air. For many midnorthern sky watchers, Sirius never climbs higher than about 30° or three fists up at culmination in the southern sky. When we gaze at the star, we're looking through not only more but denser air because our gaze is more horizontal than vertical. More air, more cells and more possibilities for turbulence.

If you really want to see Sirius go nuts, make a point of watching it when it's just a few degrees above the horizon. Then, your gaze crosses hundreds of miles of the lower atmosphere rich in air pockets of various sizes and temperatures on the move like water purling in a stream. Sirius responds with explosive-looking flashes, and if you pay close attention, you'll notice the blasts come in all the colors of the rainbow.

We learned as kids that white light is woven of every color of the rainbow, each of which is bent or refracted to a different degree. Blue is refracted more than green, which is refracted more than yellow, which is . . . on down to red. A passing eddy of air might send a red ray our way one moment, while diverting the blues and greens in another direction. The next moment, a blue ray comes our direction as red is shuttled off in another direction. Speed it up in real time and Sirius looks like a miniature pyrotechnic display with everything but the noise.

When you make your twinkle expedition, bring binoculars. They'll help brighten and intensify the colors even more. Binoculars or not, spend at least a few minutes with the star not only to witness the fluttery color show but to catch a rare moment when the world's brightest star completely vanishes.

I'm serious. Sometimes a random pocket will refract nearly *all* the light away from your gaze, even if only for a moment, and Sirius practically disappears. Now that's some awesome atmospheric trickery.

How to see Sirius twinkle best

I've listed a few times below when the star is low in the southeastern sky and most likely to be twinkling during convenient evening viewing hours. Weather conditions vary greatly, with stars twinkling less overall when viewed over an ocean versus the land, but near the horizon is always a good bet no matter where you happen to be.

Windy conditions in the wake of a cold front can produce a lot of turbulence as can times when the jet stream passes overhead. Your TV weatherperson will often show the position of the jet stream on a map. If you see it over your region, consider it an invitation to go out for a look.

To find Sirius, shoot an arrow through Orion's Belt, the famous "three stars in a row" asterism that comes up in the southeastern sky on November evenings, downward to the left. The first bright star it pierces will be the brilliant white gem, Sirius.

Best twinkle opportunities:

- Mid-November around 11 p.m.
- Mid-December around 9 p.m.
- Mid-January around 7 p.m.

Don't feel bound to these times. If you make a habit of glancing up at Sirius on clear winter nights, you're bound to catch it winking at you.

RESOURCES

- Stellarium or a phone app like Sky Chart described on page 10
- National Weather Service local forecast: weatherservice.co/us/1/?keyword=national+weather+service+forecast

Supernova

We're long overdue. Not since 1680 has anyone witnessed a supernova explosion in the Milky Way Galaxy. That August, English astronomer John Flamsteed inadvertently recorded a star he labeled as 3 Cassiopeia in the familiar W of Cassiopeia. At the time, it was a faint point of light right at the naked eye limit. Later observers saw nothing there, scratched their heads and assumed he was mistaken. Then, in 1948, a strong source of radio energy and the remains of a supernova were found very close to the position noted by Flamsteed. Today we know the ballooned-out remnant of the blast as Cassiopeia A.

Before that, in 1604, humanity got its last look at a truly bright supernova. Called Kepler's Star because it was studied by the famous astronomer Johannes Kepler, it became as bright as Jupiter. The new star blazed into view on October 9, 1604 in southern Ophiuchus in the southwestern sky and remained visible for weeks.

⋏ *One of the best candidates to explode as a supernova is the bright red star Betelgeuse in the constellation Orion, seen here rising above the tree line at left. Credit: Bob King*

Using data from the European Integral satellite in the early 2000s, astronomers estimated that a massive star should explode as a supernova once every 50 years in our galaxy. So why have we had to chill for more than 300 years?

Nobody knows for certain, but cosmic dust may be to blame. Debris shed by stars in both peaceful and violent ways hangs like filigree curtains across the light-years. Where the dust is thick enough, a star "behind the curtain" can be dimmed by many magnitudes. Perhaps this may explain in part why a Milky Way supernova is so long in coming. They've been out there but squelched by dust.

That's certainly the case with G1.9+0.3, a supernova remnant like Cassiopeia A, located close to the center of the galaxy and utterly obscured by thick clouds of interstellar dust. Astronomers estimate the star went supernova about 140 years ago. Were it not for dust, the good people of the Victorian era would have marveled at the object.

Supernovae are the brightest, most spectacular explosions in the universe. Most involve two very different kinds of stars—supergiants dozens of times more massive than the Sun and tiny, planet-size white dwarfs. Stars in the middle like the Sun are not massive enough to explode.

All stars shine because they "burn" or fuse simpler elements like hydrogen and convert them into more complex ones like helium, carbon and oxygen, with energy in the form of heat and light as byproducts. The heat and outward pressure from burning pushes back at gravity, preventing it from crushing the star like a ball of crumpled paper in your palm. During a star's lifetime, pressure and gravity strike a balance, producing stable stars like our Sun.

Massive stars have hotter cores and burn their fuel faster, eating through hydrogen, helium, carbon, neon and half a dozen other gases until iron is produced. Because a star can't burn iron as fuel, its stellar furnace shuts down. With no heat and pressure to beat back gravity, it implodes. Material rushes to the core, compressing it into a tiny, city-size remnant called a **neutron star**, then rebounds in an expanding shock wave that tears the star to bits in a titanic blast visible halfway across the universe. Quite the way to call it a day.

Big blasts, called **Type II Supernova**, aren't just for the big boys. Tiny, superdense white dwarf stars, what remains after Sun-like stars blow off their outer layers (page 174) can also explode under the right circumstances. Some white dwarfs live single lives, but others closely orbit a companion star in a binary system. So close that the dwarf's gravitational pull siphons away gas from its companion. Over time, the dwarf gains weight. This wouldn't be a problem except that nature limits a white dwarf's maximum weight to 1.4 solar masses. Add more and a catastrophic gravitational collapse follows. Runaway fusion races through the shrinking dwarf, burning and destroying it in another gigantic kaboom called a **Type Ia supernova**.

Both types occur in all sorts of galaxies. Amateurs and professionals use robotic telescopes to photograph thousands of galaxies every clear night and compare them to earlier images in hopes of finding new, exploding stars. They've done well. In 2016, over 6,400 extragalactic supernovae were discovered with more than 4,000 in the first half of 2017. Some of them reached magnitude 11.5, bright enough to see in a 6-inch (150-mm) telescope.

If you're serious about seeing a supernova in your lifetime, your best bet is to stay in touch with the flood of supernova discoveries until you find one bright enough to see in a modest telescope. Set the bar at magnitude 12, a comfortable limit for a 6-inch (150-mm) telescope from a reasonably dark sky, and visit David Bishop's Latest Supernovae website. In the column of recent discoveries running along the left side of the page, you'll find the supernova's name, e.g., 2017 eaw, 2018 B and so on. Click the link to get the particulars about the star including its host galaxy and the brightness, distance and direction from the galaxy's center. Distances are given in **seconds of arc**: 60 arc seconds = 0.75 the apparent size of Jupiter in a telescope. For reference, a 6-inch (150-mm) telescope will separate two stars separated by just under 1 arc second.

Then, use a program like Stellarium or a mobile phone app to locate the galaxy, aim your telescope there on the next clear night and use the photos posted on the site to help you pinpoint the object. This is obviously a project for those of you with good knowledge of the constellations and familiarity with using those stars as steppingstones to galaxies.

The rest of us will need the patience of ants. Although blue-white and white supergiants are massive enough to explode as supernovae, the most likely candidates are red supergiants and the blue variable star, Eta Carinae, which we met earlier. Antares in Scorpius the Scorpion and Betelgeuse in Orion are by far the brightest and easiest-to-see red supergiants. Betelgeuse is expected to go boom sometime in the next 100,000 years, with Antares bowing out within the longer span of 300,000 years.

At their distances of 650 and 550 light-years, respectively, each will become nearly as bright as the full moon. But while the Moon spreads its light over a generous half-degree-wide disk, Betelgeuse and Antares will pack it into a point, one so bright it will cast sharp, dark shadows and light the night much like the Moon.

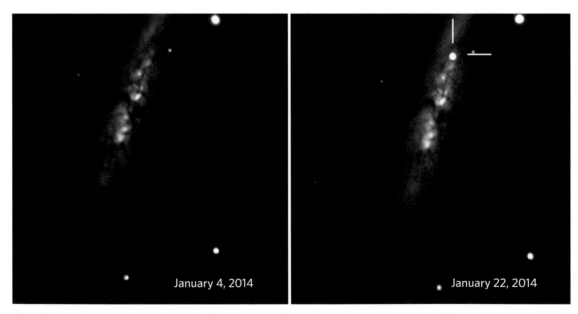

▲ *Supernova 2014J in M82, the Cigar Galaxy, first appeared in mid-January 2014 and reached a peak magnitude of 10.5, bright enough to spot in a 4-inch (100-mm) telescope. These before and after images capture the dramatic act. Credit: Scott MacNeill, Frosty Drew Observatory*

Because of their distance, neither star poses a serious threat to Earth. The gamma rays shot off by a *nearby* supernova could damage the planet's atmosphere by destroying the ozone layer, thereby exposing ocean and land-dwelling creatures to dangerous amounts of ultraviolet light from the Sun, but we're fortunate there are no nearby supergiants.

What a pity should either star be erased from the firmament. Orion would lose his right shoulder while the Scorpion would do worse and lose its heart.

How to see a supernova

If it's the kind you need a telescope for, your wait time will be only a year or two at most. Make a regular check of David Bishop's site below for supernovae brighter than about magnitude 12. Bright ones usually pop off in bright, nearby galaxies, making the job of finding one easier than you might think.

As far as Betelgeuse, Antares and Eta Carinae, there's no telling. Any could self-destruct as soon as the next clear night. Use the charts provided to keep an eye on them.

RESOURCES

- David Bishop's Latest Supernovae: www.supernova.thistlethwaites.com/snimages/

- List of supernova candidates: en.wikipedia.org/wiki/List_of_supernova_candidates

46

Nova

A supernova's a big bada boom. The star is either completely destroyed in the blast or a tiny remnant core (called a **neutron star**) or black hole remains. A nova's a less catastrophic affair. No star is destroyed in the process, but the explosion can boost an obscure star's brightness to more than 100,000 times that of the Sun.

Nova means "new" in Latin and refers to what appears to be a new star in the sky, one never seen before. There's a double irony here in that the star's actually been there all along and novae involve white dwarfs, which are some of the oldest stars in the universe.

Utility companies burn natural gas and coal to produce energy that streams through power lines to your computer and lamplights. Stars "burn" or fuse hydrogen by squishing hydrogen atoms in their cores together to make helium. The Sun burns 600 million tons (544 million metric tonnes) of hydrogen a second to make 596 tons of helium "ash" and 4 tons of energy that powers the star and warms the Earth.

Tons of energy—how is that? Solid stuff is just a super-compressed form of energy the way Albert Einstein first described it in his famous equation $E=mc^2$. Because "c" equals the speed of light or 186,000 miles per second—a large number—when you square c and multiply it times the mass (m), you can see that even a single Skittle possessed an intimidating amount of energy.

A white dwarf begins its life as a sun-like star burning hydrogen in its core, but as it ages, it not only switches to burning helium but starts burning hydrogen outside the core, causing the star to expand into a bloated version of its former self called a red giant. Helium burning produces carbon and oxygen "ash," material which a sun-like star can't use for fuel. The core shrinks and heats up as pulses of heat from burning outside the core push away the star's outer shell into space, exposing the super-hot, super-dense core now called a **white dwarf**. The process is wonderfully reminiscent of metamorphosis in the insect world, where a larva transforms to a pupa and ultimately becomes a butterfly.

A white dwarf needs a little help to get to the nova stage. It must be in close orbit with another star like the Sun or perhaps a red giant that's on the path to becoming a white dwarf. This cozy relationship provides the dwarf with a new fuel source. Here's how it works.

Even though the dwarf is only the size of Earth, it's so incredibly compressed that one cubic inch weighs 11,000 tons (10,000 metric tonnes). Orbiting close to its companion, the dwarf pulls hydrogen gas from the star into a spinning disk that looks like a glowing pancake with the dwarf at the center. Gas along the disk's inner edge gets funneled down to the star's surface.

Over time, it gets heated and compressed by the dwarf's gravity until its temperature climbs to about 36 million °F (20 million °C), about as hot as the center of the Sun. You can guess what happens next. Like tossing a match at gasoline, the hydrogen ignites in a brilliant fusion explosion. The material burns up, and we see the temporary conflagration as a "new star" back here on Earth.

The star has always been there, thousands of light-years away, but never noticed before because it's been well-behaved for millennia. But when the dwarf sets fire to its pilfered hydrogen, the inferno causes the star to brighten by many magnitudes and suddenly stand out in the night sky, looking exactly like a new star. What we're really seeing is a colossal hydrogen bomb going off!

In most novae, the detonated material gets blasted into space at several hundred thousand miles per second and creates a temporary expanding fireball of glowing debris. While the explosion is huge, except in rare instances, the star keeps it together and continues to siphon away hydrogen from its companion only to explode again thousands of years in the future.

Fifty or more stars are thought to "go nova" in the Milky Way each year, but from Earth's vantage point, we see just 5 to 10 annually. Most of those require a small telescope, but about once a year, a nova will become bright enough to show either in binoculars or with the naked eye. The brightest in recent times was Nova Delphini 2013 that blew its top in August 2013 and shot up to magnitude 4.4. It was easy to see with the naked eye from suburban areas and lots of fun to follow as the star surged to maximum and gradually faded. Five years later, only larger amateur telescopes still showed it—a dim spark 14,700 light-years away.

As the exploding debris cloud expands, it cools, causing hydrogen to give off red light, a color

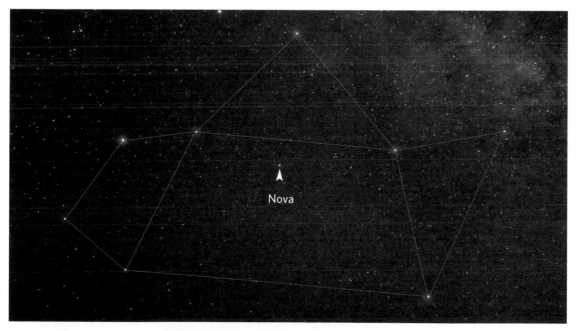

▲   *In March 2015, a new bright nova appeared in the Teapot of Sagittarius in the pre-dawn sky and peaked at magnitude 4.3, easy to see with the naked eye if you knew just where to look. It was the brightest nova to appear in that constellation since at least 1898. Credit: Bob King*

you can easily see in a small telescope. Looking at a nova and understanding what you're seeing stirs an inner vibe. For me, it's an opportunity to witness how the universe works. Every clear night, the cosmos offers up amazing sights and demonstrations of basic scientific principles you might only otherwise read about.

Novae can occur anywhere in the sky, but most happen along the band of the Milky Way because that's where most of the stars in the galaxy are concentrated. More stars mean a better chance that one of them will pop as a nova. Where stars are sparse, say in Ursa Major the Great Bear, the sample's too small. Not a single nova's been recorded there in modern history.

The brightest nova in more than 80 years shot into view in Cygnus the Swan, better known as the Northern Cross, in August 1975. It peaked at magnitude 1.7, so all you had to do was look up, and there it was. Located three "fingers" above the head of the Cross, I still recall how the "new star" distorted the outline of the constellation, a sight that's stayed with me a lifetime.

How to see a nova

Novae are discovered every year, but bright ones are uncommon. To find out when the next one will appear, you'll need to subscribe to a couple of free email services that send alerts about new discoveries. Two are listed below. When using The Astronomer's Telegram (ATel) service, go the site, subscribe and then select your areas of interest in the left-hand column. Some of the subjects are technical, some obvious. If you only want news of novae, click the nova checkbox. Every time a nova is discovered, including extremely faint ones in other galaxies, you'll get an alert. In reading the alerts, watch for the words "early outburst" or "caught on the rise." These are the ones to pay attention to.

The VSNET service will send you updates of unusual activity in a variety of different kinds of stars. Most of the ATels and VSNET alerts you can just delete, but be on the lookout for bright outbursts.

When a nova is discovered, its sky coordinates are given. Akin to latitude and longitude but called **declination** (Dec) and **right ascension** (RA), they precisely locate the star in the sky. You can take these numbers and plunk them into either an interactive online star map such as the one listed below or type them into one of several free star map software programs such as Stellarium (open the app, click the magnifying class, then select the Position tab and type in the RA and Dec) and the program will pinpoint its location in the sky.

RESOURCES

- The Astronomer's Telegram: astronomerstelegram.org/
- VSNET alert: ooruri.kusastro.kyoto-u.ac.jp/mailman/listinfo/vsnet-alert
- American Association of Variable Star Observers (AAVSO), A clearinghouse for the latest on variable stars including novae: aavso.org
- Antelope Valley Astronomy Club interactive star chart: avastronomyclub.org/skymap/d/skymap.php
- *Stellarium* or a phone app like *Sky Chart* described on page 10.

All of Those Other Satellites

47

The last thing you want to hear from a boyfriend, girlfriend or spouse are the dreaded words, "I think we should start seeing other people." The news can be heartbreaking. Nothing like that will ever happen with satellites. If you've had success finding the International Space Station (page 98) using online resources or apps, you have all the skills you need to look for and identify other satellites. Why restrict yourself to just one? I'll bet you're just as curious as I am about what else is up there.

⋏   *The Lacrosse 5 rocket stage became as bright as Deneb in the Northern Cross as it sped north near Arcturus (bottom left) during a pass in July 2017. The rocket helped place a military spy satellite in orbit in 2005. Credit: Bob King*

As of early 2017, there were 1,459 active satellites in orbit and roughly 7,500 inactive ones, including everything from the final rocket stages that went along for the ride into Earth orbit, to defunct reconnaissance, science, communications, GPS satellites and associated parts and pieces. And those are just the big ones. While the U.S. Strategic Command tracked 17,852 objects in orbit in July 2016, it's estimated there are 170 million pieces of debris smaller than a centimeter (July 2013) whizzing around up there.

Some of the stranger things that ended up in orbit before ultimately burning up in Earth's atmosphere were a glove lost by astronaut Ed White during his 1965 Gemini 4 flight; a portion of the ashes of Gene Roddenberry, creator of *Star Trek*; and a tool bag that slipped from Heide Stefanyshyn-Piper's grasp while she worked on a space station solar panel in 2008. Classified as ISS DEB (TOOL BAG) and given the official NORAD ID #33442, it was easy to see in binoculars before its orbit finally decayed and the bag met a fiery end.

A fair number of these rocket stages and other satellite-associated hardware tumble as they orbit the Earth because they're uncontrolled. Their metallic surfaces act as mirrors and reflect sunlight in regular or random patterns toward the observer. Satellite aficionados call them flashers. Others, such as Japan's Experimental Geodetic Payload (EGP), are designed to twinkle. Its 7-foot (2.2-m) sphere is covered in mirrors and reflectors that make it sparkle like a strobe light when viewed through binoculars.

To begin working the unplowed ground of lesser-known but easy-to-spot satellites, I compiled a list of suggestions from other satellite watchers and my own observations, which you'll find at the end of this entry. To find them, I suggest you use the now familiar Heavens Above site, the same as we discussed earlier, for tracking the space station and Iridium satellites (page 102).

While there are other great online satellite tracking sites and software prediction programs, I like Heavens Above because of its easy-to-use options and helpful maps. To start, sign in and select your town, then return to the homepage and click on the Daily Predictions for Brighter Satellites link along the left side. You'll be taken to a list of satellites you can filter by magnitude depending on how dark your sky is.

I select magnitude 4 (underneath the drop-down boxes for the date) for my minimum brightness, which pulls up a list of dozens of manmade stars for evening viewing that will reach magnitude 4 or brighter. If you're out in the wee hours, select the Morning button. Magnitude 4 works very well for outer suburban and rural skies, but if you live closer to a bigger city, narrow the list further to magnitude 3.

Even though I can see down to magnitude 6 from my observing site on a moonless night, fourth magnitude is plenty dim for naked-eye satellite tracking. Be aware that the listed brightness is the peak magnitude of the object—for much of its path, it can appear considerably fainter. That's the main difference between the ISS and other satellites. We've gotten so spoiled with its brilliance from start to finish, it's a no-brainer following an entire pass. Every other satellite is much smaller than the ISS, and many travel in higher orbits, so we can only follow them across a segment of sky, about one-quarter to one-third of their full path, before they fade from view.

Clicking the satellite's name will call up a map of its path with minute-by-minute positions marked along that path. Maps are extremely helpful as it's crucial to know just where to look to anticipate a favorite rocket stage's arrival.

Next, prepare a list of good candidates for the time you plan to observe and either have the website handy on your mobile phone or write/print a short description of location and direction of motion. Then, head out and enjoy an evening of satellite watching, knowing that some of these birds—spy satellites—may be watching you, too. Be aware you may run into an occasional no-show. This is typical. I just move on to the next one on my list.

Want more details about what you're seeing? Click on the Info link at the upper right on the map page.

You'll soon discover that many of the objects are the upper stages of rockets used to send a myriad of Russian Kosmos-series satellites into orbit. These include military reconnaissance, science and lunar probes. NASA and ESA are no slouches, either, when it comes to providing rocket bodies for viewing.

Orbits 101

The majority of working satellites and all crewed space stations are in low-Earth orbit (LEO) and range in altitude from about 111 miles (180 km) to 1,243 miles (2,000 km) above Earth. This includes the ISS, Hubble Space Telescope, Earth observation and spy satellites and the Iridiums. Next most numerous are more than 400 television, communications and weather satellites in geosynchronous orbit (GEO). These satellites orbit at greater than 22,300 miles (36,000 km) altitude and have orbital periods of 24 hours, the same as Earth's rotation. They "hover" over the same location and provide either a continuous stream of photos of the same region of the planet or serve to relay TV signals around the globe.

A smaller number of satellites, including the global positioning satellites (GPS) so helpful for finding the nearest ice cream shop, and space environment satellites for measuring radiation effects and space debris, occupy medium-Earth orbits (MEO) at altitudes from about 1,243 miles (2,000 km) to 22,300 miles (36,000 km).

All satellites receive a 5-digit NORAD catalog number, which makes it very handy to look any of them up on Heavens-Above. To use this function, return to the Heavens-Above homepage and click on the Satellite Database link. In the box, fill in the NORAD (Spacetrack) number, check the In Earth Orbit Only box, tap enter, and you'll see the satellite at the top of the list. Next, click the Visible Passes link and you're all set.

Because most satellites don't come close to the brilliance of the ISS, make a point to observe them either when the Moon is in a "small" phase and not very bright or when it's completely out of the sky.

One of the things you'll enjoy about adding new birds to your list is seeing with your own eyes the very satellites that provide crucial data or take beautiful photos such as the Hubble Space Telescope. The Earth-sensing Terra and Aqua satellites, both easy naked eye targets, gather information about the planet to help us better understand climate change. You'll also have first-hand experience with space junk, forever a hot topic in the news. Once you're comfortable spotting new satellites, invite friends and family over and share a little bit of space-age history. Everybody loves these artificial stars.

Here's a list of satellites to cut your teeth on:

- Lacrosse 5 R/B (rocket body)
- Atlas-Centaur R/B (multiple rocket bodies in orbit)
- Terra
- Cosmos R/B (multiple rocket bodies in orbit)
- BREEZE-M Debris Tank
- SL-16 R/B (multiple rocket bodies in orbit)
- Tiangong-1 and Tiangong-2 (Chinese prototype space stations, both as bright as magnitude 1.5)
- The pair of TerraSar-X and Tandem-X
- Hubble Space Telescope/HST (for observers in the southern United States)
- ERS-1
- Aqua (Terra's counterpart)
- Envisat
- H2-A R/B (rocket body from November 2009 launch of IGS-Optical 3 satellite)
- Cosmo-Skymed 1 (#31598)
- USA 267 (#41334)
- USA 215 (#37162)
- Okean O (#25860)

How to see other satellites

Use the detailed directions given above in combination with Heavens-Above and the other resources listed below. Because the Earth's shadow doesn't cover as much of the sky in general during the summer compared to the winter, you'll see more satellites crisscrossing the sky in the warmer months compared to the colder. Most convenient!

RESOURCES

- The Seesat-l satellite watchers mailing list. Stay in touch with all the satellite news and latest software: mailman.satobs.org/mailman/listinfo/seesat-l

- Heavens-Above: heavens-above.com

- Visual Satellite Observer's Homepage. A nice, easy guide to satellites: satobs.org

- N2YO.com. Satellite tracking site. It automatically recognizes your location. You can then use it to find pass times and ground tracks for more than 18,780 objects. This is a great site to use if the NORAD number doesn't yield results at Heavens-Above. Just put the number in the Find a Satellite box and then click 10-day-predictions at left.

- Space-Track.org. Create an account to download the latest orbital elements for use in free satellite tracking software programs: space-track.org/auth/login

- Heavensat. Popular satellite tracking software: sat.belastro.net/heavensat.ru/english/index.html

- List of 100 (or so) brightest satellites: celestrak.com/NORAD/elements/visual.txt

(Note: Parts of this chapter have appeared in *Sky & Telescope* and are reprinted with permission.)

48

Stars on Water

Science is an uncharted river that winds on and on. Dip your paddle in the water and get ready for perspective-changing adventures around every bend.

Paddle-dipping is something I enjoy whenever I get the chance and a calm lake is an ideal place for stargazing. On a still night from a canoe or kayak, you can look up and take in the vastness of the starry sky, then look down and see those same points of light glimmering in black water. Is this what it feels like to be an astronaut? Free of Earth—suspended between the stars?

Any movement of the water, any ripple will cause the stellar reflections to stretch and pulse, in contrast to the steady stars above. It's about quietude. Shutting off the noise. Many of us find it

∧ *Floating at the boundary of starry sky and starlit lake gives us an opportunity to ease back and take stock.*
Credit: Bob King

crucial to clearing the mind and regaining our footing on a planet where so much happens so fast. Floating at the interface of distant suns and their wet reflections, the noise dies down and we find ourselves reconnecting with the natural world. If the stars do nothing else, they make us pause. That's good enough.

How many ways are there to enjoy a starry sky? Probably as many as there are stars in the sky. Standing on a hilltop, from the middle of a cornfield, a car window or your own front yard, each offers a unique perspective from which to appreciate the sky.

Earth is an observatory, after all. A place from which we look out into the universe, ponder our purpose and consider the possibilities. It seems as if we know so much, and compared to our ancestors from 5000 B.C., we do. But imagine how basic our knowledge will seem to our descendants in the year 5000. In science, we have to accept a partial and ongoing understanding of the world.

Despite science's tentative nature, we know of no better tool for making sense of the world and getting answers about the how and why of life and the universe. Nature generously reveals its secrets to the curious mind, and the knowledge gained only deepens our sense of wonder.

Where to float your boat

Pick a time when there's no Moon in the sky and a place where there's little to no light pollution. Lakes without too much outdoor cabin lighting are understandably the best places. Even a pond will do. Many state and national forests and other federally protected areas have plenty of lakes from which to choose. Minnesota's Boundary Waters Canoe Area Wilderness or Florida's Everglades National Park make ideal places to visit on vacation, but there are many others.

Don't forget to wear a life jacket and bring a headlamp to negotiate getting into and out of your craft.

RESOURCES

- Fourteen best places to canoe or kayak in national forests: nationalforests.org/blog/fourteen-best-places-to-canoe-and-kayak-on-national-forests
- Boundary Waters Canoe Area Wilderness reservations: bit.ly/2v2Vrsi

49

Crab Nebula

When we looked at the possibility of seeing a supernova in our lifetimes, there were two options. You could either track one down in another galaxy, which requires a telescope, or wait for Betelgeuse or another of the supergiant stars to explode. The latter is always a possibility but hopelessly beyond our control.

There is another option: Go to the scene of a recent, bright supernova that occurred in our galaxy and look for its remains. I know just the place. Let me introduce you to the Crab Nebula, located near the tip of the southern horn in the constellation of Taurus the Bull.

On July 4, 1054, Chinese astronomers recorded a brilliant new star at this spot that quickly grew six to ten times brighter than the planet Venus. For more than three weeks it was visible in the daytime sky and remained a naked-eye sight until April 6, 1056—a total of more than 21 months.

▲   *This Anasazi painting on a cliff underhang in Chaco Culture National Historical Park in New Mexico may depict the supernova of 1054 AD and the nearby crescent moon at the time the explosion occurred. Credit: Alex Marentes/CC 2.0*

There's evidence the stellar explosion may have been recorded by an Anasazi Indian artist in a pictograph on a rock overhang in Chaco Canyon in New Mexico. I hiked to the spot many years ago, during one of my first astronomical pilgrimages, and took pictures of the ten-pointed star and crescent moon. From there, the artist would have had a wide-open view to the east, the very direction where the supernova first appeared in the dawn sky.

Using Stellarium or similar sky mapping software, anyone can now take a trip back in time to the morning of the star's discovery and see for themselves how the waning crescent moon joined the scene on July 5; it must have been an incredible sight. While no one can definitively prove the pictograph records that long-ago morning scene, the circumstantial evidence is strong. Archaeologists know for a fact that the Anasazi inhabited the area at the time, and the painting is a good representation of the event. And yet, it could be something else, perhaps a conjunction of the Moon and Venus. We'll never know.

But the fact remains that something extraordinary happened that July morning. What was it? For that, we have a clear answer: A supergiant star some eight to ten times more massive than the Sun located about 6,500 light-years from Earth went supernova.

⌃   *This composite image of the Crab Nebula was taken by 5 different observatories in radio, infrared, visible light, ultraviolet, X-rays and gamma rays. The white dot at center is the Crab pulsar. Source: NASA, ESA, G. Dubner (IAFE, CONICET-University of Buenos Aires) et al.; A. Loll et al.; T. Temim et al.; F. Seward et al.; VLA/ NRAO/AUI/NSF; Chandra/CXC; Spitzer/JPL-Caltech; XMM-Newton/ESA; and Hubble/STScI*

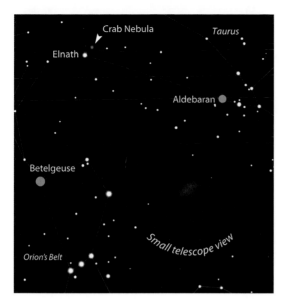

The Crab is easy to find in a small telescope 1° northwest of the 3rd magnitude star, El Nath, in Taurus. Source: Stellarium

Like the supernovae we've touched on earlier, the former star ran out of fuel. With its internal furnace now shut down, gravity took hold and the stellar giant underwent a sudden collapse, imploding and then exploding in a powerful supernova blast. The detonation created an expanding, glowing cloud of debris—the Crab Nebula—and a tiny, supercompressed remnant of the original star's core called a **neutron star**.

The star is only 12 miles (19 km) across—about the size of a small city—and rotates 33 times a second, shooting out beams or pulses of radiation at precise intervals just like a lighthouse. When a pulse sweeps over the Earth, we see a faint point of light. When it moves on, the star goes dark until the next beam sweeps by. One follows another so quickly, the star appears to flicker. A beaming neutron star is called a **pulsar**.

The nebula's intricate shape formed through the interplay of the pulsar's light beams, its wind of fast-moving particles and the expanding debris cloud of the original explosion.

While the pulsar's too faint to spot in even larger amateur telescopes, the glowing, expanding blast cloud is bright enough to show in a pair of 7x50 binoculars under a dark sky. Appropriately, it looks like a little puff of smoke. A 6-inch (150-mm) telescope reveals an elongated, textured cloud that seems to float against the background stars like a ghost.

The Crab and others of its kind are known as supernovae remnants. There are several in the sky, but it's the only bright remnant that astronomers can tie to a witnessed, recorded event. It also serves to link us to our ancestors who nearly a thousand years ago witnessed the lighting of this candle. For us, only the smoke still lingers.

How to see the Crab Nebula

The Crab is well-placed for viewing in the evening sky from November through March and easily found using Betelgeuse in Orion and Aldebaran, the Bull's Eye. In the late fall, Taurus appears in the eastern sky at nightfall; from early to midwinter, due south and above the upraised club of Orion; and in late winter, standing in the west.

Using the map, connect the third-magnitude star, El Nath (Zeta Tauri), into a triangle with Betelgeuse and Aldebaran. Point your telescope or binoculars at the star and look just about 1° to its northwest to spot the supernova's remains. You can observe it at any magnification, but I recommend 75x to 100x.

50

Barnard's Star

Stars are constantly on the move. Or at least it looks that way. Every clear evening, you can watch them climb up in the eastern sky and then slide down toward the western horizon. Earth rotates *toward* the east, so not only do the stars rise in that direction but the Sun, Moon and planets, too.

Ancient skywatchers thought the stars were attached to a celestial sphere that revolved around the fixed, immovable Earth once a day. For nearly 500 years now, we've known it's really the other way around. Earth's doing the spinning and making the stars *appear* to move across the sky.

Do stars move on their own apart from Earth's rotation? Yes! But seeing it with your eyes is quite another thing. Each of the stars in the night sky has its own particular motion, but centuries must pass to see even the smallest shift in their positions with the naked eye.

Around 135 BC, the Greek astronomer Hipparchus made careful measurements of the positions of more than 800 stars and compiled the first comprehensive star catalog. More than 1,850 years later, English astronomer Edmund Halley compared the positions of the stars in the year 1717 with those listed by Hipparchus and got a surprise: two of them—Arcturus and Sirius—had moved. Not by

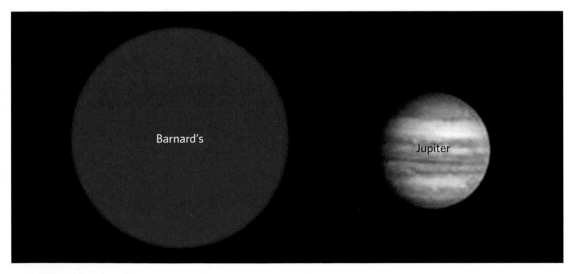

⌃ *Barnard's Star, a fast-moving red dwarf in Ophiuchus just 6 light-years from Earth, measures 1.9 times Jupiter's diameter. Source: Wikimedia with additions by the author*

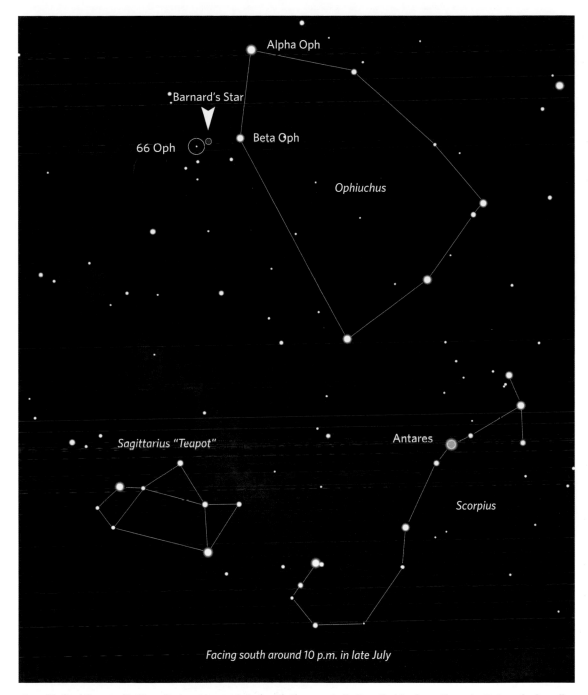

To find Barnard's Star, first get acquainted with the constellation Ophiuchus, the celestial snake handler, located above the scorpion and teapot constellations. Barnard's lies 4° east of the 3rd-magnitude star Beta Ophiuchi and less than 1° northwest of 4.5 magnitude 66 Ophiuchi. Once you find 66 Oph, switch to the detailed map. Source: Stellarium

much, but enough that a keen-eyed skywatcher from the days of ancient Greece transported to the present would notice. Between the time of Hipparchus and Halley, Arcturus had moved 1° (or twice the diameter of the full moon) to the southwest and Sirius a smidge more than 0.5° (one Moon diameter) in the same direction. Halley had discovered the true motion of stars called **proper motion**.

Stars, including the Sun and the solar system, move around the great mass of stars at the center of the Milky Way Galaxy. Even traveling at 514,000 mph (828,000 kmh) it takes the Sun 230 million years to complete an orbit. That incredible length of time gives us a taste of how big a place the galaxy is. Other stars in the night sky are moving at similar breakneck speeds, but not a single one visible with the naked eye will appear to budge in our lifetime. The reason? They're all much too far away. Think of a distant airplane crossing the sky. It might be cruising at 550 mph (885 kmh), but to your eye, it appears to creep along. The same is true of the stars except that hundreds of years must pass for us to notice their movements.

Halley caught Arcturus in the act because it's moving at 272,000 mph (439,000 kmh) nearly perpendicular to our line of sight, so it only takes about 1,000 years for a reasonably astute skywatcher to notice its flight. Because no one has that kind of time on their hands, we'll have to look elsewhere to detect stellar motion. Fortunately, a number of other fainter stars have higher proper motions. One of the easiest to spot in a small telescope is Barnard's Star in the constellation Ophiuchus the Serpent Bearer.

Just six light-years from Earth, Barnard's the second closest star to Earth after the Alpha Centauri system. Discovered by American astronomer E. E. Barnard in 1916, it scoots faster across the sky than any other star in the heavens, covering 0.25° or half a full Moon diameter in a human lifetime.

That's plenty fast for anyone with a 4-inch (101-mm) or larger telescope to detect its northward movement in a year or two. Astronomy teaches patience if nothing else. Although the star is not difficult to find and see, it will take some experience at the telescope to pencil this must-see

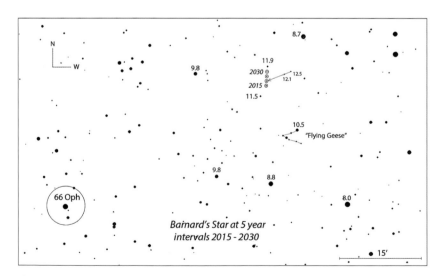

▲ *Once you've found 66 Oph, use this map to star-hop west and north to Barnard's Star, whose positions are marked every 5 years. Source: Chris Marriott's Sky Map with additions by the author*

onto your bucket list. Go after Barnard's Star after you've found bigger, brighter objects like the Hercules Globular or Ring Nebula.

First, get familiar with the big, bell-shaped outline of Ophiuchus, a constellation located above Scorpius the Scorpion in the Summer sky. Locate the third-magnitude star Beta Ophiuchi (shown on the map) with your naked eye, then point your telescope 4° east (to the left) to fifth-magnitude 66 Ophiuchi. Once at 66, use the detailed map to star-hop from there to Barnard's Star just 1° to the northwest.

With a magnification of 75x to 150x, make a sketch of the star in relation to the stars near it. Compare Barnard's position to the magnitude 11.9 and 11.5 stars, or watch for it to form a straight line with a pair of twelfth-magnitude stars to its northwest between now and 2020.

A clipboard with a blank sheet of unruled paper and a red flashlight—to help you see what you are doing—makes for an ideal setup. Take note of Barnard's Star's color. It's red because the star is a red dwarf star just 1.9 times the diameter of Jupiter. Red dwarfs are much fainter and cooler than the Sun, so although the star is only 1.6 light-years farther from us than brilliant Alpha Centauri, it shines at magnitude 9.5, right at the limit of a standard pair of binoculars.

The next step is a simple one: Forget about the star for an entire year. Better yet, stay away for *two years*. Stow away your sketch and seek out some of the other must-sees. When you return, compare your sketch to where the star is now, and you'll see that it has really moved. Sketch the star's new position and stop by for another visit in the future.

Seeing a star move in your lifetime lets you break through the illusion that the sky is static, that stars are fixed. All is in flux as suns whirl around the galactic core. Over the course of tens of thousands of years, the individual motions of the stars will alter the outlines of the present constellations, creating new ones for our distant descendants.

How to see Barnard's Star

Ophiuchus and Barnard's Star are well placed for viewing in the evening sky from June through September. You'll need a small telescope and moderate (75 to 100x) magnification for the best views. Plan on observing the star when the Moon's not too bright, so you can include the fainter stars around it in your drawing. You'll need them to track its travels.

The close-up map will help you pinpoint 66 Ophiuchi. Once you acquire it in your telescope, use the fainter stars on the detailed map as stepping-stones to Barnard's Star.

RESOURCES

- The discovery of stellar proper motion: narit.or.th/en/files/2010JAHHlvol13/2010JAHH...13..149B.pdf

- More about Barnard's Star: earthsky.org/astronomy-essentials/barnards-star-closest-stars-famous-stars

51

Eta Aquilae, Heartbeat Star

Watching a star physically change in just a week seems too much to expect. On the contrary. The star Eta Aquilae (AY-tuh AK-will-lie), located just down the block from the bright star Altair in the Summer Triangle, brightens into plain view, fades away and then rebrightens every 7.17 days. And it does so with the precision of a fine watch.

Eta's bright enough to follow from the light-polluted skies of the outer suburbs with a peak magnitude of 3.5, about a level fainter than the Big Dipper stars. I always make sure to glance in its direction when I'm out for a stroll on summer and fall nights. Doing so keeps me in touch with its simple but compelling rhythm. Once you've identified the star, it takes only a little effort to follow its ups and downs without optical aid.

When you spot the star for the first time, there's no telling where it will be on its curve: maximum, minimum (when it fades to magnitude 4.3) or somewhere in between. Just watch it for a few nights and you'll get the drift soon enough. An entire cycle unfolds in just a week's time, making this one of the easiest variable stars to follow. If you miss it one week because of clouds, just drop by the next.

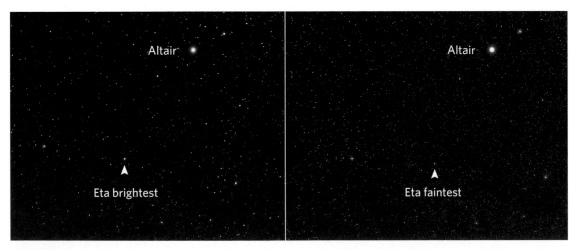

▲   *Eta Aquilae's brightness swings are easy to spot with the naked eye. The star takes a little more than a week to complete a cycle. These photos were taken three nights apart in October 2017, when the star faded from maximum to near minimum. Credit: Bob King*

Eta Aquilae is a yellow-white supergiant 65 times larger than the Sun—and 3,400 times more luminous—located about 1,400 light-years from your doorstep. What a shame it's not a little closer or it would be one of the brightest suns in the sky. Eta is a **Cepheid** (SEF-ee-id) variable star, named after the prototype Delta Cephei in the constellation Cepheus the King.

Cepheids are unstable stars that pulsate regularly, like beating hearts. They shrink and expand with periods that range from 1 to 100 days, causing their temperature and brightness to vary in a very regular way. Most Cepheids rise rapidly to maximum light followed by a slower fall to minimum.

Eta is brightest when it's expanding like a blown-up balloon and faintest when it's contracting like a balloon losing air. Thankfully, the sun doesn't seesaw back and forth like this or temperatures on Earth would alternate between roasting and freezing every week.

Most stars visible on a clear night burn hydrogen as fuel to create energy and lead relatively stable lives. Not Cepheids. After exhausting the hydrogen in its core, a Cepheid burns the next element up the chain—helium—while simultaneously burning hydrogen in a shell just outside the core. For a time, it becomes unstable and begins to pulsate.

While the pulsations have much to do with how changing temperatures affect helium in the star's atmosphere, we can boil it down to this: A Cepheid expands and overshoots a stable size until gravity pulls it back in again. As it shrinks, the push-back pressure from the compacting gases slows and stops the collapse. When the pressure exceeds gravity's tug, the star expands and begins another pulsation cycle.

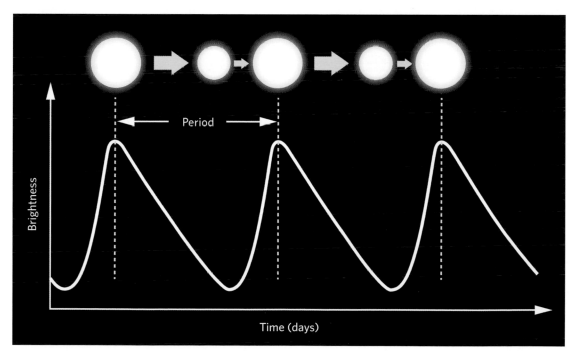

▲ *Cepheid variable stars like Eta Aql pulsate in size and brightness with a regular period. Credit: Gary Meader*

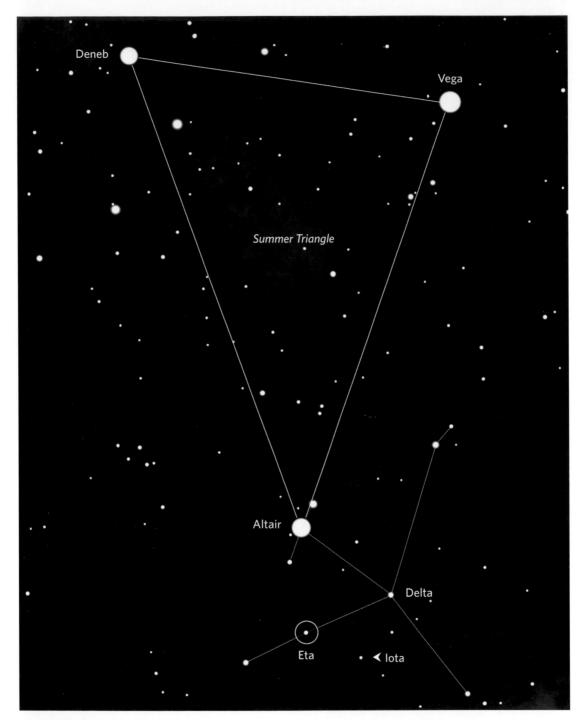

∧ *To find Eta Aql, start with the bright star Altair in the bottom of the Summer Triangle—Eta's about a fist due south. At peak brightness (magnitude 3.5), it's similar to Delta Aql; at minimum (4.3), it's a match for Iota Aql. Source: Stellarium*

Eta Aquilae's ups and downs were first recognized by the sharp-eyed English amateur astronomer Edward Pigott in 1784. It's one of the brightest Cepheids known and ideal for beginning skywatchers to get acquainted with the quick-turnaround light cycles of these heartbeat stars.

How to find and watch Eta Aquilae

Use the finder map to help track down the star, located 8° (not quite one fist) below and left of Altair, the bottom star in the Summer Triangle. I've labeled stars near Eta with magnitudes, so you can more easily compare its changing brightness. Aquila is well-placed in the evening sky from July through November.

RESOURCES

- Pulsating variable stars. A nice introduction: www.atnf.csiro.au/outreach/education/senior/astrophysics/variable_pulsating.html

- Cepheid variables. Go here for a more technical but still understandable explanation of pulsations: cseligman.com/text/stars/variables.htm

52

Reddest Star

In astronomy, seeing red can be a good thing. Most stars look white to the naked eye because they're not bright enough to stimulate the cone cells in our retina that sense color. Only the brightest stars like Antares, Betelgeuse and Spica betray their hues. A telescope gathers much more light than the unaided eye, enough to pump up the brightness of otherwise faint, "colorless" stars into full color. Anyone using a scope will be impressed by the burning purity of stellar colors that range from golden yellow to ruby red.

Some of the most esthetically beautiful suns belong to a class called **carbon stars**. Most carbon stars are red giants. The outer envelope or "surface" of a red giant is cooler than that of a hotter star like the Sun or Sirius and glows orange or red.

But carbon stars take color to the next level. We learned earlier that as stars age they switch from burning hydrogen to helium. Burning helium creates carbon "ash." Convective currents, like the ones that bubble up in a pot of boiling water on the stove, dredge carbon from the core and deliver it to the star's outer layers. There, it condenses into a fine soot that absorbs blue and green light. Only oranges and reds penetrate the dusty barrier to reach our eyes, making carbon stars some of the reddest in the sky.

⌃ *The variable R Leporis, one of the reddest stars in the sky, has great eye appeal. Credit: Joseph Brimacombe*

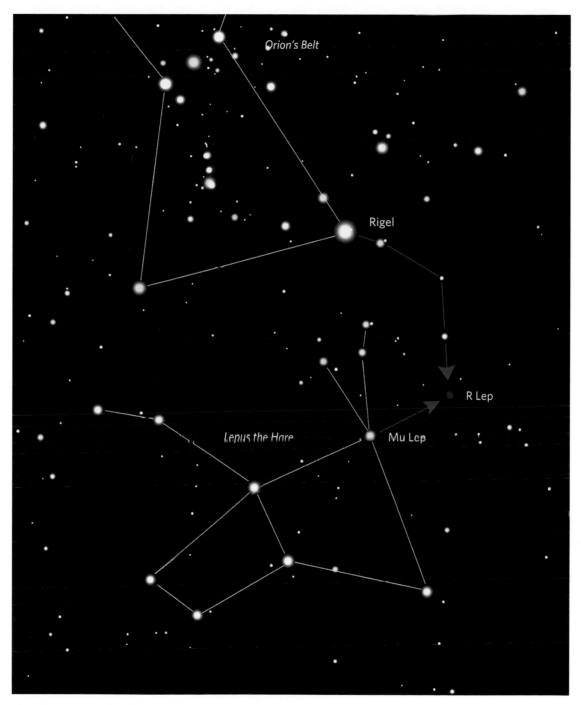

▲ *You can use Rigel, Orion's brightest star, or Mu Leporis to point you to R Lep in the constellation Lepus the Hare, located directly below Orion. Source: Stellarium*

One of the most striking red stars is R Lep, short for R Leporis, in the constellation Lepus the Hare. Lepus crouches under Orion the Hunter and comes into good view every December. To find R Lep, start at Mu Leporis and star-hop 3.5° west as shown on the map. The star varies in brightness from magnitude 5.5, bright enough to see dimly with the naked eye, down to a faint 11.7. Most of the time it's between 6 and 10 and within easy reach of a 3-inch (76-mm) telescope.

All carbon stars are variable stars that vary in brightness with periods ranging from a couple months to more than a year. Their color also depends on other more subjective factors including how bright the star is at the time (the brighter, the more washed out the color), who's doing the looking (you see colors a bit differently than I do) and the Purkinje Effect. The last depends on how long you *stare* at a red star. If you happen to catch R Lep at the faint end of its cycle, it will appear less colorful at a glance than if you stare at it continuously.

R Lep is also known as Hind's Crimson Star after its discoverer J.R. Hind, a nineteenth-century English astronomer who stumbled on this stellar ruby in 1845. Many observers describe its color as intense, smoky red or compare it to a drop of blood, but it always reminds me of the shiny reflection from a red Christmas ball ornament, an association that springs from the star's return to the evening sky during the winter holiday season.

R Lep's light varies over 432 days, and if you catch it near the bottom of its cycle between magnitude 8 and 10, it will look more intensely red than at the bright end. The natural perfume of a flower, like the intense sweetness of a white clover blossom, captivates your nose the same way the sight of a deeply-reddened R Lep makes it hard to look away.

How to see R Lep

The constellation Lepus is visible during convenient evening viewing hours from December through March. To find it, first locate the Belt of Orion. Look about one fist below and right of the Belt to find the brilliant white star Rigel in Orion's Foot. Now, look about one fist directly below (south) of Rigel to find the third-magnitude Mu Lep. From here, use the map to star hop to R with your telescope. If you have any difficulty, just look for a colorful red star at the location—the smoking gun.

RESOURCES

- American Association of Variable Star Observers (AAVSO). Check on the star's current brightness by typing "R Lep" in the Pick a Star box, then drop two lines down and click the Check Recent Observations link : aavso.org/

- If you fall head over heels for red carbon stars, check out the Astronomical League's Carbon Star Observing Program at: astroleague.org/content/carbon-star-observing-club and its list of 100 carbon stars at: astroleague.org/files/obsclubs/CarbonStar/CarbonStarList3.pdf

53

Mercury Transit

About thirteen or fourteen times a century, the planet Mercury passes directly in front of the Sun exactly the way the Moon does during a total solar eclipse. Because the Moon's so much closer, it completely covers the Sun. Mercury, in contrast, with a diameter of just 3,032 miles (4,879 km), faces an impossible task. Although a bit bigger than the Moon, it's 48 million miles (77 million km) from the Earth and only covers a tiny bit of the Sun. Instead of an eclipse, we see a *transit*.

I saw my first transit in May 1970 after toting my telescope through the woods to the edge of a field in northern Wisconsin, where the Sun stood in the clear. I remember the sight of that perfectly round black dot creeping across the Sun's face. I haven't missed a transit since 1973, 1999, 2006 and 2016. Three involved adventure, whether that was dropping in unannounced at an amateur astronomer's observatory in a small town in Austria (1973) or outracing clouds to find clear skies in 1999 and 2006. In 2016, for once, neither last-minute travel nor cloud-anxiety played a part. I simply got up, drove to our local planetarium and shared the sight with amateurs and passersby under sunny skies.

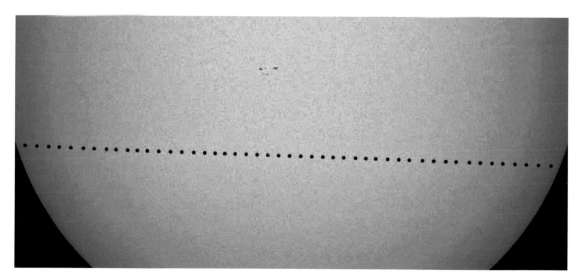

➤ *This composite image, taken with NASA's Solar Dynamics Observatory, shows Mercury's progress across the Sun's face during the May 9, 2016 transit. Source: NASA/SDO*

Adventures are common in the astronomy hobby. When you're not fighting clouds, the clock or needing to travel, you look around a bit dumbfounded. How can it be this easy?

Mercury's orbit is tilted 7° to the plane of Earth's orbit, more than twice as much as any other planet, so the only time we get to see it cross the Sun's face is when it happens to lie in the exact same direction of the Sun at the same time intersects Earth's orbital plane. This happens about seven times a century. Because some of those transits happen when the Sun's up on the other side of the world, most of us will see only a few in a lifetime.

Transits vary in length depending on how centrally Mercury's path crosses the Sun. Most last several hours, so there's a good chance you'll see at least part of one through a break in the clouds. But sometimes the planet barely grazes the Sun's edge or crosses only a short distance inside the solar limb. The November 15, 1999 transit finished up in just 52 minutes. When the planet crosses near the center of the solar disk, it also happens to be closest to the Sun—and therefore farthest from the Earth—and appears to move more slowly. The 1970 event lasted nearly eight hours.

So why stare at a black dot? First, we get a sense of the Sun's enormous size. You can't appreciate this any other way except during a transit, when it's possible to approximately compare a planet's girth with that of the Sun. Mercury looks tiny indeed. Granted, it's the smallest planet, but if Earth were put in its place, its silhouette would be only 2.6 times larger.

Transits also provide the opportunity to visualize a planet's location in three dimensions. Mercury's average distance from the Sun is 36 million miles (57.9 million km), so when you see the dot, picture it hovering in the foreground about 57 million miles (91.7 million km) from your nose.

Watching a transit is also the most popular method for detecting exoplanets, the name given to planets that orbit other stars. As of this writing, nearly 4,000 have been found, many using the **transit method**. When a planet transits a star, it temporarily dims it by a small but measurable amount. Repeated dips in one star's light can tell us a lot about the planet, including its size.

Scientifically, astronomers use Mercury transits to measure changes, if any, in the diameter of the Sun, compare drops in the light levels between solar and stellar transits and investigate the "black-drop effect." The black-drop effect occurs *at* and moments *after* the planet contacts the Sun's inner edge (contact II—see next page) and again at the inner edge on the Sun's opposite side (contact III). Both times, the planet's disk looks slightly stretched out as if it can't quite "let go" of the Sun's inner edge on the way in or reaching out to touch it on the way out. Several factors including diffraction, atmospheric turbulence and less-than-perfect telescope optics are the cause.

Only Mercury and Venus have transits because they orbit *between* the Earth and Sun. All of the other planets lie beyond Earth's orbit; when they pass near the Sun in the sky, they're on its opposite or far side. Venus transits are even more amazing than Mercury's because the planet is 2.6 times larger and considerably closer to Earth. With a proper solar filter, Venus is plainly visible to the naked eye as a striking black disk.

Venus transits occur in pairs eight years apart separated by long gaps of 121½ years and 105½ years. The last pair occurred in June 2004 and June 2012. I'd be jumping up and down about seeing the next ones were it not for the fact that it won't happen until December 10 to 11, 2117 and December 8, 2125. Be sure to tell your kids' grandkids to watch.

How to see a Mercury transit

Transits always occur in either November or May when Mercury is in conjunction with the Sun while crossing the Earth's orbital plane at the same time.

All that's needed to view one is decent weather and a small telescope equipped with a safe solar filter. Mercury's silhouette is too small to see with the naked eye. Use the transit list below to plan your outing.

A transit has four stages: contact I, the instant Mercury touches the outer edge of the Sun; contact II, when it touches the inner edge; contact III, when the planet touches the Sun's inner edge again moments before departing the disk; and contact IV, when Mercury last touches the outer edge of the Sun.

Times of upcoming Mercury transits

Times are in Eastern Standard for November transits and Eastern Daylight for May events except when a transit isn't visible in North America. Then, Universal Times are given.

NOVEMBER 11, 2019

- Start: 7:35 a.m. Middle: 10:20 a.m. End: 1:04 p.m.

NOVEMBER 13, 2032 (NOT VISIBLE IN NORTH AMERICA EXCEPT EXTREME EASTERN CANADA BUT IDEAL FOR THE EASTERN HEMISPHERE)

- Start: 6:41 UT. Middle: 8:54 UT. End: 11:07 UT

NOVEMBER 7, 2039 (NOT VISIBLE IN NORTH AMERICA)

- Start: 7:17 UT. Middle: 8:46 UT. End: 10:15 UT

MAY 7, 2049

- Start: 7:03 a.m. Middle: 10:24 a.m. End: 1:44 p.m.

RESOURCES

- NASA Eclipse website—planetary transits across the Sun: eclipse.gsfc.nasa.gov/transit/transit.html

54

Orion's Belt

Thank goodness for asterisms. Without them, finding the constellations would be a lot harder. An asterism can be either a bright part of a constellation or parts of two different constellations connected together to form a distinctive or familiar pattern. Orion's Belt, comprised of three equally bright stars in a row, is the best known of them. Next best? The Big Dipper, of course.

▲ Orion's Belt, the best-known asterism in the night sky, peeks out from between silhouetted trees on a winter night. Credit: Bob King

Let's take a minute to explore the Belt. The trio of stars makes its first appearance in the evening sky in mid-October when it rises in the east around midnight. A month later, with Daylight Saving Time out of the way, it's up by 9 p.m., and at Christmas, commands the southern sky.

Besides their pretty symmetry and compactness, the Belt stars stand out because they're all relatively bright, each shining at around second magnitude. Less well known are their names. Alnitak (ALL-nit-ahk) on the Belt's east side, Mintaka (MIN-ta-ka), on the west—both mean "belt" in Arabic—and the center star, Alnilam (ALL-nil-ahm), represents a "string of pearls." String of Pearls is also the Arabic name for the entire belt. All three are massive stars 90,000 to 375,000 times more luminous than the Sun.

Like the constellations, many asterisms are happenstance alignments of stars at wildly different distances. While the Belt stars aren't related, they all lie moderately far from Earth in the same nook of the galaxy: 800 light-years for Alnitak; 1,340 light-years for Alnilam and 915 light-years for Mintaka.

The Belt goes by many names both in current times and across human history: The Three Kings, The Three Sisters, the Three Marys (after the Marys in the Bible), Jacob's Rod, Peter's Staff, the Magi and the Yardstick. Whatever way you choose to see it, Orion's Belt is the key to the city. With it, you can connect the rest of the "dots" around it to make the outline of the constellation Orion the Hunter. Look one fist above the Belt to spy the bright red supergiant star Betelgeuse in the hunter's right shoulder, then over to Bellatrix, the left arm, down to shimmering Rigel in his foot and over to Saiph, which marks his left knee. Together, these four stars form a big box around the Belt.

If your skies allow, don't stop there. His sword dangles below the Belt, while curls of fainter stars that arc from Betelgeuse and Bellatrix outline a club and lion's pelt, also depicted as a shield. Above the Betelgeuse–Bellatrix line, look for a small triangle of stars that form a puny head.

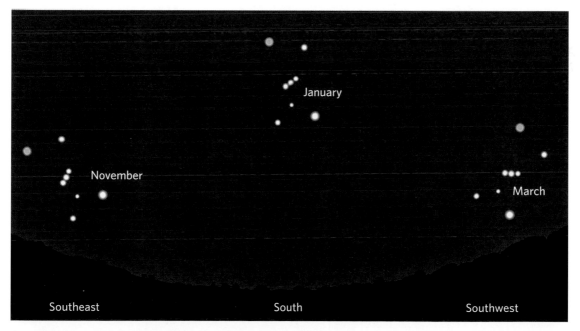

▲ *This map will help you keep track of Orion's seasonal ride across the southern sky. Positions are shown for mid-month around 10 p.m. local time. Source: Stellarium*

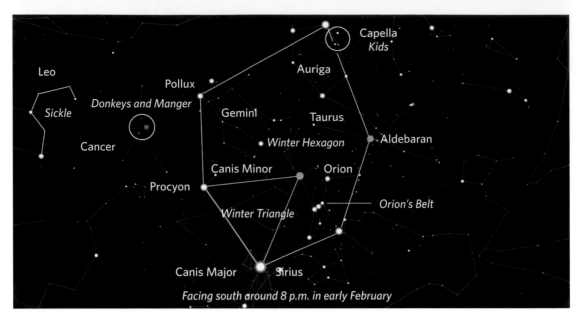

▲ Orion's Belt isn't the only asterism in the winter sky. The brightest stars in the south starting with Sirius (bottom) connect to form the enormous Winter Hexagon, part of which doubles as the Winter Triangle. Leo's

▼ Summer's loaded with delightful asterisms, all of which are keys to finding that season's constellations. They include a huge diamond, the Summer Triangle, Hercules' Keystone and more. Source: Stellarium

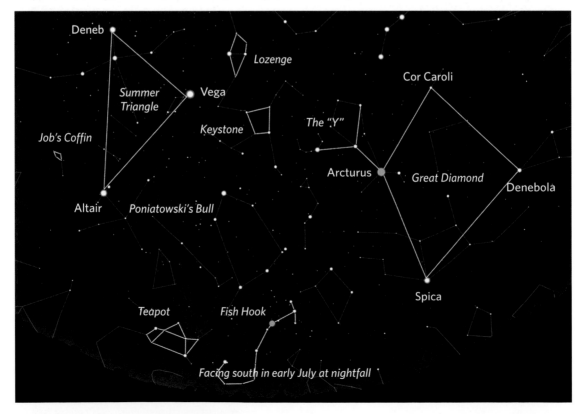

Asterisms are everywhere. The Big Dipper will guide you to Ursa Major the Great Bear (see page 31); the Summer Triangle to the constellations Cygnus, Aquila and Lyra; the Teapot to Sagittarius; and the Great Square to Pegasus and a half-dozen different constellations.

There are even telescopic asterisms, some with imaginative names like the Golf Putter, Napoleon's Hat and the Sunken Crouton. If naked eye asterisms help to identify constellations, telescopic ones are useful for tracking down star clusters, nebulae and other deep-sky objects. They're shorthand ways for an observer to move from pattern to pattern until the target is reached. We're all born experts at seeing and creating patterns out of random information. If you've ever imagined a face or animal among the turrets and folds of a puffy, summer's-day cloud, you're up to asterism-making as you learn the landscapes of the night sky.

How to see Orion's Belt and other asterisms

Use the map provided to find Orion's Belt and the rest of the constellation or fire up a star map app on your mobile phone or computer. Orion is easily visible in the evening sky from November through March and in the pre-dawn sky from August through October.

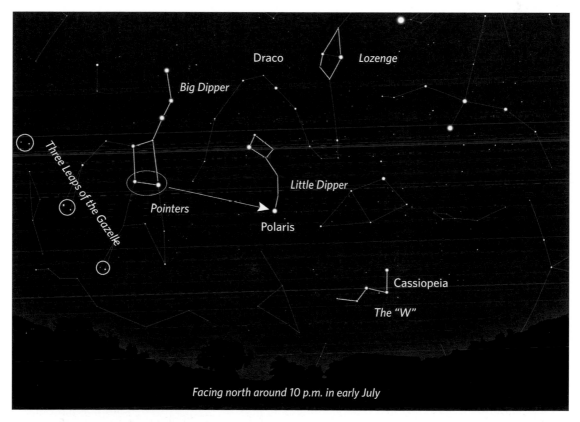

Facing north around 10 p.m. in early July

▲ *Besides familiar Dippers and W of Cassiopeia, the northern sky has several lesser known patterns you'll want to seek out. Source: Stellarium*

55

Ursa Major Moving Cluster

The Big Dipper is more than what it seems. The familiar pattern isn't just a chance alignment of stars like so many other asterisms and constellations but rather the core of the nearest star cluster, named the Ursa Major Moving Cluster. The Hyades cluster in Taurus, the one shaped like the letter "V" that hosts the bright red-orange star, Aldebaran, is the closest star cluster. You'll find it about a fist to the left of the Pleiades or Seven Sisters cluster in the late fall and winter sky. The Hyades are 151 light-years away and contain several hundred stars.

The Ursa Major Moving Cluster (UMMC) is a much looser, sparser group of only thirteen or fourteen stars almost half as close or about 78 light-years from Earth. It's barely a cluster, the reason you'll see it sometimes described as a "group," but all its stars are related as they are in more familiar clusters. They were born together as siblings in a great cloud of gas and dust roughly 500 million years ago and now occupy a volume of space some 30 light-years long by 18 light-years wide.

If you glance at the Big Dipper, you're seeing its core and brightest members. All but two of the Dipper stars belong; Alkaid at the end of the handle and Dubhe, the topmost star in the bowl, are unrelated to the cluster and only appear in the same line of sight. Several fainter stars in Ursa Major and one star in the neighboring constellation, Canes Venatici the Hunting Dogs, comprise the cluster. There are also 42 additional stars broadly scattered across the sky from Cetus the Sea Monster to Lupus the Wolf that belong to a larger "stream" of stars related to and possibly members of the cluster.

How do we know the stars of the UMMC are related? In 1869, English astronomer Richard Proctor discovered they were all moving together through space, drifting slowly in the direction of the constellation Sagittarius at the rate of one full-Moon diameter every 16,000 years. Finding stars sharing the same motion and age is typically how astronomers identify related stars from random ones.

Our solar system is located on the outskirts of the cluster, but its much greater age makes it unrelated. Meanwhile, as we plant tomatoes and shovel snow, this sparse collection of suns is moving in our direction at 7 miles per second (11.5 km a second). In about a million years, it will pass us at a distance of about 51 light-years.

You might wonder if the Sun was ever part of a cluster. It may well have been, but most star clusters like the Hyades and UMMC have relatively short lifetimes of a few hundred million years. That may sound like practically forever but not in relation to a typical star like the Sun, which will live to be more than 10 billion years old before running out of fuel.

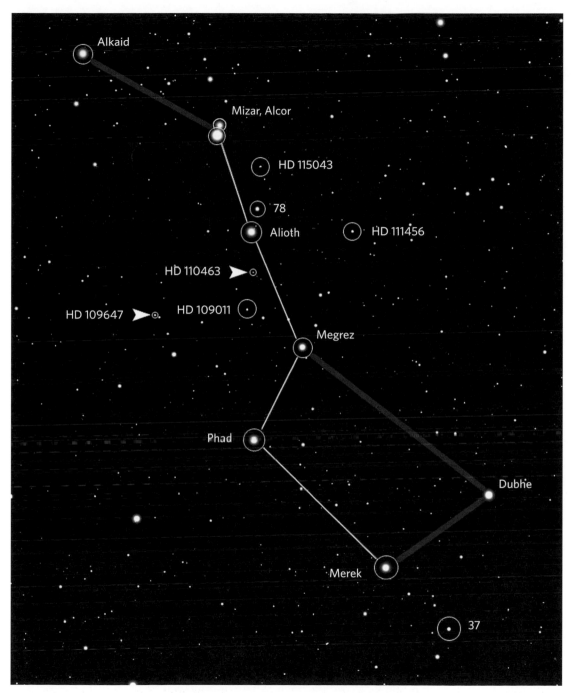

▲ *One of the Big Dipper's little-known secrets is that most of its stars form the core of a star cluster called the Ursa Major Moving Cluster. Source: Stellarium with additions by the author*

Unlike the beehive-like balls of globular clusters, stars in Hyades-type clusters, called **open clusters**, are only loosely bound by their mutual gravities. A star's individual motion can cause it to break away and leave its siblings behind. Star clusters are also subject to bigger forces intent on ripping them apart. As a cluster revolves about the galactic center, massive clouds of dust and gas as well as tides produced by the shear mass of stars in the Milky Way take it apart star by star.

Still, with the right tools, we might try to identify which far-flung stars may once have been the Sun's brothers and sisters. Stars born together share similarities in their atmospheric chemistry and motions through space. Astronomers have studied a number of nearby stars and identified at least one good candidate, a star slightly more massive than the Sun called HD 162826, 110 light years away in the constellation Hercules. They may have once been pals.

There are some sights in the night sky that may not appear compelling at first blush but once you know what you're seeing, they help us appreciate them on a deeper level. Something as familiar as the Dipper can take us on a journey across many light-and clock-years, informing us of the grand movements of stars that twinkle delicately overhead.

How to see the Ursa Major Moving Cluster

The core of the cluster in the Big Dipper is easy to spot most of the year in the northern sky but best from late winter through summer, when the Dipper stands high. Use the accompanying map to identify other members of the cluster, nine of which are visible with the naked eye and four that require binoculars.

RESOURCES

URSA MAJOR MOVING CLUSTER:
- ottawa-rasc.ca/astronotes/1998/an9803p4.html

SUN'S SIBLING FOUND:
- skyandtelescope.com/astronomy-news/sun-sibling-found/

- Stellarium or a phone app like Sky Chart described on page 10

56

Top Ten Lunar Wonders

Anyone with a pair of binoculars—or better, a small telescope—all are there in profusion. Craters, mountain peaks and ranges, lava flows, cliffs and clefts. Because the Moon is so close, you can see details smaller than a mile (1.6 km) across in a 6-inch (150-mm) telescope. After working the planets hard to glimpse a cloud feature or vague dark markings, the Moon gives so much for so little effort. Just point and look, and if the seeing conditions are reasonably tranquil, you'll soon be transfixed.

Besides Mars, the Moon is the only other body in the solar system with a surface we can see clearly. The Apollo leftovers including flags, rovers and descent modules are well beyond the capability of even the Hubble Space Telescope, but at least we can see where they landed.

When first-time scope users get a look at the Moon, I hear at least as many "wows" as when they see Saturn. How to boil down hundreds of sights to a few must-sees. In choosing these, I may face criticism from some long-time moonwatchers for leaving out a favorite, but I guarantee to get you off to a good start.

You already know the Moon goes through phases: new moon, evening crescent, half—called **first-quarter phase**—waxing gibbous, full, waning gibbous, last or third quarter, morning crescent and back to new. When the Moon waxes from evening crescent to full, the advancing line of *sunrise* on the Moon, called the **terminator**, moves to the east, exposing a new slice of the Moon's surface each night. After full phase, the terminator defines the advancing line of *sunset*. Each night, more and more of the Moon becomes covered in shadow and removed from view.

If you were a lunar astronaut standing somewhere along the terminator, you'd see the Sun at sunrise or sunset very low on the horizon. As on Earth, a low Sun casts long shadows. Shadows reveal lots of detail on the Moon's surface the same way shadowy light clearly exposes the lines and textures in a person's face. Every hill, crater wall and cliff on the Moon throws a shadow when the terminator is nearby, making it the best place to see lunar features under the most dramatic lighting.

I've listed and described ten of the finest sights on the next page, along with the number of days after the new or full moon when they look most spectacular. We'll be following the advancing terminator as we move from one highlight to the next:

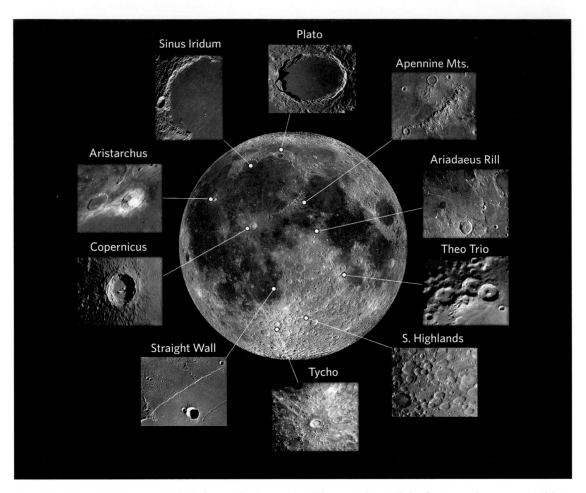

▲ *Use this mini Moon guide to help you find our Top 10 lunar sights. All the features shown are visible in a small telescope and some in binoculars. Source: NASA/GSFC/Arizona State University /Damian Peach / Bob King*

Theo trio—Theophilus, Cyrillus and Catharina—An eye-catching trio of larger craters each about 60 miles (95 km) across and around 2 miles (3.3 km) deep. You can easily see that Theophilus, the one with the obvious central mountain peak, is the youngest because it looks crisper and overlies the older, more eroded crater, Cyrillus.
Best seen: 6 days past new moon or 4 to 5 days past full moon

Ariadaeus Rill—A long, narrow crack resembling a valley. Some rills were once conduits for molten lava, but this one formed when a section of the Moon's crust between two faults sunk to create a segmented valley 145 miles (233 km) long. It's located between the Southern Highlands and Apennine Mountains.
Best seen: 7 to 8 days past new or 5 to 6 days past full

Southern Highlands—Crater-mania! Feast your eyes on this region of ancient lunar crust that still bears the scars of thousands of overlapping impact craters of all sizes. You can guesstimate their relative ages by comparing those that have sharper rims to others that are softer and more eroded.
Best seen: 7 to 9 days past new or 5 to 7 days after full

Apennine Mountains—A rugged arc of mountains with peaks reaching altitudes of 1.2 to 3 miles (2 to 5 km) named after the Apennines in Italy. The range is relatively young at 3.9 billion years compared to the 4.4 billion-year-old highlands. Unlike Earth's mountain ranges, which form through movements of crustal plates, the lunar Apennines formed in the aftermath of a large asteroid impact that excavated the neighboring Sea of Showers (Mare Imbrium) basin. The curving shape of the mountains forms part of the rim of the basin.
Best seen: 8 to 9 days after new moon or 6 to 7 days after full

Copernicus crater—One of the most magnificent lunar craters especially when observed straddling the terminator with its interior filled to the brim with inky black shadow. Fifty-eight miles (93 km) across and 2.4 miles (3.8 km) deep, Copernicus displays several small central peaks, terraced interior walls caused by slumping and, at full moon, a corona of bright rays. These rays formed when material from the impact rained back down and pocked the surroundings with thousands of secondary impacts, excavating "fresh" material beneath the lunar topsoil. The rays are best visible at full moon.
Best seen: 9 to 10 days after new or 8 days after full

Plato crater—68 miles (109 km) across, Plato's relatively smooth dark floor makes it look like an oval swimming pool. The funny description contains a grain of truth: Plato's interior was flooded with lava sometime after its formation 3.8 billion years ago. With a depth of just 0.6 mile (1 km), it's shallow compared to many other craters.
Best seen: same time frame as Copernicus

Tycho crater—When near the terminator, this 53-mile (86-km) wide diameter crater looks crisp, fresh and deep; its floor sits 3 miles (4.8 km) below the average surface level and boasts a pointy peak. One of the youngest, largest lunar craters, the asteroid that blasted out Tycho struck the Moon only about 108 million years ago, an event the dinosaurs may even have noticed. Around full moon, when sunlight shines straight down on the crater, it transforms into a brilliant white spot surrounded by an extensive "aura" of brilliant rays.
Best seen: 9 to 10 days after new, at full moon and 7 days after full

Straight Wall—A nearly linear fault 68 miles (110 km) long and about 1.5 miles (2 to 3 km) wide. Although only 787 to 984 feet high (240 to 300 m), when viewed near the waxing moon's terminator, the wall casts a prominent shadow that makes it look quite steep. After full moon, it looks like a thin, bright line.
Best seen: 9 days after new moon or 7 days past full

Sinus Iridum—Latin for Bay of Rainbows, this 147-mile (236-km)-wide lava plain is one of the Moon's most beautiful features. Two rugged promontories or arms reach out to frame a bay that in low sunlight appears filled with long, ocean-style "rollers." The waves are long ridges that formed when basaltic lavas cooled and contracted after flooding the area. In fact, Sinus Iridum is really what's left of a large impact crater with half its outline submerged beneath ancient lava flows. *Best seen: 11 days after new moon or 9 days after full*

Aristarchus crater—Only 25 miles (40 km) wide, this crater stands out better than many twice its size because it's one of the brightest spots on the Moon. It doesn't hurt that it also sits on a vast, dark lava plain called the Ocean of Storms, making for a strong contrast. Around full moon, it looks like a searchlight in the dark; bright rays make it even more prominent. Aristarchus and Copernicus are visible without optical aid as bright, diffuse spots at full moon. *Best seen: at or around full moon*

How to see lunar highlights

All you need to know is the Moon's current phase. You can get that from the *Old Farmer's Almanac*, a Moon calendar, or an app or program. From there, you can figure out the best nights—using the "Best seen" information in the list above for viewing a particular feature.

I recommend at least a 3-inch (75-mm) scope for Moon observing with magnifications ranging from 40x to 150x. Higher magnifications will make you feel like you're in lunar orbit as well as help sort out the busy Southern Highlands crater scene. As with any telescopic viewing, be sure to allow your instrument to cool to the air temperature before using it otherwise the images may be blurry for a time.

RESOURCES

- Moon phases calendar: timeanddate.com/moon/phases/
- Moon Phase Plus for iPhone (free): itunes.apple.com/us/app/moon-phase-plus/id671352640?mt=8
- Phases of the Moon for Android (free): play.google.com/store/apps/details?id=com.universetoday.moon.free&hl=en
- Virtual Moon Atlas (free). A superb, interactive moon atlas for your laptop: sourceforge.net/projects/virtualmoon/

57

Galactic Duo, M81 and M82

We've knocked around the Milky Way, stuck our noses in the Magellanic clouds, and took a trip to the Andromeda Galaxy. What lies beyond? Only about two trillion more galaxies. That's the most recent estimate of how many other Milky Ways populate the universe. To grasp the magnitude of this number, we look to time. One million seconds equal 11.5 days; a billion seconds 31.75 years and a trillion seconds 31,710 years. Two trillion seconds comes to 63,420 years. Tick . . . tick . . . tick . . . tick . . . one second for each galaxy.

We can use binoculars to leap beyond Andromeda to a pair of galaxies not far from the bowl of the Big Dipper nicknamed Bode's Galaxy and the Cigar Galaxy. Both were cataloged by Messier as M81 and M82 but discovered by German astronomer Johann Bode on December 31, 1774. They're bright and easy to see as galaxies go, with magnitudes of 6.9 and 8.4, respectively, and they are favorite targets for beginning and amateur astronomers.

▲ *The bright pair of galaxies, M81 (left) and M82 are visible in 50-mm binoculars under reasonably dark skies. A few observers have even seen them without optical aid. Credit: Markus Schopfer/CC 2.5 Generic*

Through 7x35 binoculars in dark skies, they're dim side-by-side smudges, but in a pair of larger 10x50s, you can see much more. After a pleasant star-hop from the bowl in the general direction of the North Star, look for two gray patches of fuzz. They're close together, so if you find the brighter one, M81, it will take you directly to the fainter Cigar, puffing away just 0.5° to its north.

Both are easy to see in the northern sky most of the year but highest when the Big Dipper's nearly overhead each spring. Because looking straight up is hard on the neck, the view is more comfortable in summer, when you'll find them conveniently placed in the northwestern sky at nightfall.

Their apparent closeness isn't a line-of-sight ruse. Both are about 12 million light-years away or almost five times farther than Andromeda and physically close together in space, separated by just 150,000 light-years. You might even say they're in a relationship—The gravity of each has stripped gas from the other. In M82, that material has fallen into the galaxy's center, become compressed and sparked a vigorous new round of star formation.

None of this tumult is directly visible in binoculars or modest amateur telescopes, but we won't let that stop us. What you *can* see in binoculars is M81's oval shape and brighter center, where stars are concentrated in greater numbers.

The smaller and fainter Cigar lies due north of M81. If your eyes are fully dark-adapted and you use averted vision, its tilt is just discernible. Your eyes are fine instruments capable of seeing more than you think, given the chance. Take in the scene and try to appreciate that these twin fuzzies represent the combined light of some 100 billion suns, the light of which left the galaxies

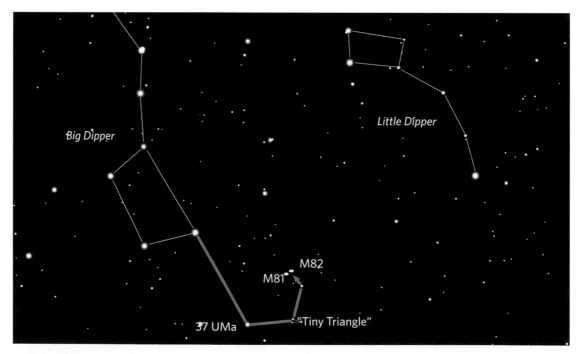

▲ *Use the Bowl of the Big Dipper to find 37 UMa and then star-hop from there past the "Tiny Triangle" to our galactic duo. Source: Stellarium*

when Earth knew no human footprints. The fact that we can see such exotica from our own homes reminds us that the universe is never far away. Sunlight shines into our houses and apartments. The Moon raises tides. Galaxies remind us how far space goes.

And it goes and goes and goes. Not far from M81 and M82, just above the back end of the Dipper's bowl, lies an unremarkable patch of sky with no obvious stars or galaxies. Here, astronomer Bob Williams of the Space Telescope Science Institute used the Hubble Space Telescope to take 342 separate exposures over ten consecutive days between December 18 and December 28, 1995. Basically, Hubble stared at *nothing* for more than 100 hours. But within this shard of sky measuring less than a tenth the diameter of the full moon, his images revealed a rich assortment of more than 3,000 spiral, elliptical and irregular galaxies, some of which date back to twelve billion years, not long after the origin of the universe in the Big Bang. Many are small, irregularly shaped and glow blue with new stars—teething babies that would later merge with other small galaxies to build many of the bigger galaxies we see around us today. Because light takes billions of years to reach our eyes from these little ones, we see them as they were billions of years ago.

Williams's use of valuable time on the Hubble scope to stare into space was scoffed at by some astronomers, but the gamble revealed that beyond the empty spaces between the stars lie billions of galaxies undreamed of only decades ago. It also provided crucial evidence showing how myriad small galaxies from the early days of the universe merged with others to form bigger ones.

Mergers continue to this day, with some surprisingly close to home. The Milky Way's two largest satellite galaxies, the Magellanic Clouds, appear to be on a collision course with the home galaxy sometime in the far future. But that's only a warm-up for the BIG MASHUP. The Andromeda Galaxy is screaming toward the Milky Way at nearly 245,000 mph (110 km/sec) and the two will merge some four billion years hence into a single galactic behemoth some are already calling Milkomeda.

How to see the M81–M82 galaxy pair

A few amateur astronomers under exceptionally dark skies have seen both galaxies without optical aid, but most of us will need a pair of binoculars. The best time to look is when the Moon is out of the sky from midwinter through late summer, when the galaxy pair stands high in the northern sky free of the dust and haze that plague viewing near the horizon. Use the map provided and 35-mm or larger binoculars (7x35, 8x40, 7x50, 10x50) or a 3-inch (75-mm) or larger telescope. You'll be looking for two small, misty patches of diffuse light 10° off the bucket of the Big Dipper. With a tripod adaptor, you can fix your binoculars to a camera tripod for super-steady viewing that makes finding fainter objects like galaxies much easier.

RESOURCES
- Hubble Deep Field and Ultra-Deep Fields: spacetelescope.org/science/deep_fields/
- Andromeda–Milky Way collision: en.wikipedia.org/wiki/Andromeda%E2%80%93Milky_Way_collision

Epilogue

Like many of you, I'm the owner of a furry Canis Major. Her name is Sammy. We always thought she was mostly border collie, but my daughter gifted me with a doggie DNA kit a few years back, and now we know with scientific certainty that she's a mix of German shepherd, Siberian husky and golden retriever. Yeah, she's a mutt.

Sammy's going on 18 years old now—that's human years—and has neither the spunk nor bladder control of a young pup. She wanders, paces, gets confused. In her aging, I see what's in store for all of us as we pass from one stage of life to the next.

Intentionally or not, we humans often leave a legacy before we depart. Maybe a big building, a work of art or an exemplary life. As I stare down at my panting dog, it occurs to me that she's leaving a legacy too, one she's completely unaware of but which I'll always appreciate.

Thanks to my dog, I've seen more auroras and lunar halos than I can count. That goes for meteors, contrails, space station passes, light pillars and moonrises, too. All of this because she needs to be walked in the early morning and again at night. This simple act ensures that, while Sammy sniffs and marks, I get to spend at least twenty minutes under the sky almost every night of the year.

I'm an amateur astronomer and keep tabs on what's up, but my dog makes sure I don't miss a beat. Let's say she keeps me honest. There's no avoiding going out, or I'll pay for it in whimpering and cleanup.

There have been times I was totally unaware that an aurora was underway until Sammy needed to be walked. When we were done, I'd often dash away to a dark sky with camera and tripod to record and share the lights. Other nights, walking the dog would alert me to a sudden clearing and the opportunity to catch a variable star on the rise or see a newly discovered comet for the first time.

Amateur astronomers routinely brush up against eternity. We observe stars and galaxies by eye and telescope that remind us of both the vastness of space and the aching expanse of time. I have only so many years left before I spend the next ten billion disassembled and strewn about like that scarecrow attacked by flying monkeys. But when I see the Andromeda Galaxy through my telescope, with its 2.5-million-year-old photons setting off tiny explosions in my retinas, I get a taste of eternity in the here and now.

That's where Sammy offers yet another pearl. Dogs seem to do a far better job living in the moment than people. They can eat the same food twice a day for a decade and relish it anew every single time. The same goes for their excitement at seeing their owner or taking a walk or a million other ways they reveal that *this* moment is what counts.

People tend to think of eternity as encompassing all of time, but Sammy has a different take. A moment fully experienced feels like it might never end. Lose yourself in the moment, and the clock stops ticking. I love that feeling. That's how my dog's been living all along. Canine wisdom: one billion years = one moment. *Both* feel like forever.

▲ *Sammy. Credit: Bob King*

Sammy's lost much of her hearing and some of her eyesight. We're not sure how much time she has. Maybe a few months, maybe even another year, but her legacy is clear. She's been a great pet and teacher, even if she never figured out how to fetch. We've hiked hard trails together and then rested atop precipices with the Sun sinking in the west. I look into her clouded eyes these days and have to speak up when I call her name, but she's been and remains a "good dog!"

Maybe you have a dog or maybe you don't, but the point is the same. The night sky lies at the junction of the infinite moment and time eternal. Open the door, step outside and look up. Infinities await.

Resources

Because a pair of binoculars and telescope are essential equipment for seeing some of the night sky wonders face-to-face, I've included some suggestions for good-quality but inexpensive equipment along with several trusted vendors.

Telescopes come in several basic designs: refractor, reflector and Schmidt–Cassegrain. The size of the mirror or lens is probably the most important consideration in buying a scope. Bigger is better to a point because a larger mirror/lens gathers more light and makes sky objects brighter and easier to see. Too big, however, can mean you'll get tired of lugging the scope outside. That's why it's best to strike a compromise between size and ease of use.

The easiest, most stable, portable and least expensive type of instrument—and what I recommend—is a Dobsonian reflecting telescope with a 4.5- to 8-inch (115- to 200-mm) mirror. You can set one up in a minute and never have to worry about wobbly, tripod-style telescopes. You also get more aperture (size of mirror) for your hard-earned dollar compared to other scope types.

No disrespect to refracting telescopes. Small ones are relatively inexpensive, portable and have large fields of view perfect for objects like the Pleiades. Schmidt–Cassegrains are short and stubby and perfect for astrophotography. One day, you might wake up a proud owner of all three!

Binoculars

- Celestron 7x50 Cometron—Inexpensive and gets great reviews. Eye relief's a bit short at 13 mm. Field of view (FOV) = 6.8° ($35)
- Nikon Aculon A211 8x42—Again, short eye relief (12 mm) but otherwise highly rated with sharp optics. FOV = 8° ($80)
- Nikon Aculon A211 7x50—A little more expensive but has a generous eye relief of 17.6 mm. FOV = 6.4° ($90 to $100)
- Canon 15x50 IS All-Weather Image Stabilized binocular—Fantastic instrument that provides bright, sharp and steady views but much more expensive than ordinary binoculars ($1,200)

Telescopes

- Celestron FirstScope 76 mm. Small, compact and a good starter scope for young children ($50)
- Orion SkyQuest XT 4.5-inch (114-mm) Classic Dobsonian reflecting telescope ($240)
- Orion SkyQuest XT 6-inch (150-mm) Classic Dobsonian reflecting telescope ($270)
- Sky-Watcher 8-inch (200-mm) Dobsonian telescope. Price is about $355 for the full-tube version and $425 for the collapsible-tube model
- Meade Infinity 80 mm altazimuth refracting telescope ($190)
- Orion StarBlast 80 mm AutoTracker refracting telescope ($180)

Eyepieces

Often, the eyepieces supplied with telescopes have rather narrow fields of view. If you'd like more of a picture-window experience instead of looking through a straw, I suggest these:

- 20 mm Orion Expanse eyepiece ($50)
- 32 mm Meade Series 4000 Super Plossl 1.25 inch (31 mm) eyepiece ($50)
- Olivon 20 mm Wide Angle Plossl eyepiece ($63)

Vendors

- B&H Photo (bhphotovideo.com)
- Orion (telescope.com)
- Opticsplanet.com
- Amazon.com
- Cloudynights.com: Amateur astronomers sell used telescopes and related equipment. Sellers are rated with feedback, so you know who you're buying from. Registration is free. The site also includes many topic-specific discussion groups well worth exploring.

Aurora Tours

- Aurora Service Tours (Finland, Lapland): tours.aurora-service.eu/or their Facebook page: facebook.com/auroraservicetours/
- For a variety of aurora tours by bus or jeep in Iceland, including those led by professional photographers: https://icelandaurora.com/photo-tours/northern-lights-photo-tours/
- Aurora Village in Yellowknife, Canada: auroravillage.com/tours-pricing
- Selection of tours in Alaska: goalaskatours.com/winteractivities_northernlights.html

Checklist

- ☐ 1. Magnificent Saturn
- ☐ 2. Moon–Planet/Planet–Planet Conjunction
- ☐ 3. Three-Planet Conjunction
- ☐ 4. Summer Milky Way
- ☐ 5. Winter Milky Way
- ☐ 6. Andromeda Galaxy
- ☐ 7. Ursa Major the Great Bear
- ☐ 8. Total Lunar Eclipse
- ☐ 9. Total Solar Eclipse
- ☐ 10. Awesome Aurora
- ☐ 11. Southern Cross
- ☐ 12. Alpha Centauri
- ☐ 13. Orion Nebula
- ☐ 14. Eta Carina Nebula
- ☐ 15. Pleiades in Binoculars
- ☐ 16. Moon Occultation
- ☐ 17. Moon Dogs
- ☐ 18. Distorted Moons
- ☐ 19. Light Pillars
- ☐ 20. Bright Meteor Shower
- ☐ 21. Jupiter's Galilean Moons
- ☐ 22. Jupiter's Great Red Spot
- ☐ 23. Mars's Polar Caps
- ☐ 24. Eight Planets in One Night
- ☐ 25. International Space Station
- ☐ 26. Iridium Flare
- ☐ 27. Vesta, Brightest Asteroid
- ☐ 28. Zodiacal Light
- ☐ 29. Best Double Stars
- ☐ 30. Ring Nebula
- ☐ 31. Perseus Double Cluster
- ☐ 32. Hercules Globular Cluster
- ☐ 33. Slender Moons
- ☐ 34. Supermoons
- ☐ 35. Luna's "Dark" Side
- ☐ 36. Uranus and Neptune
- ☐ 37. Crescent Venus
- ☐ 38. Cygnus Star Cloud
- ☐ 39. Magellanic Clouds
- ☐ 40. Magical Mira
- ☐ 41. Bright Comet
- ☐ 42. Airglow
- ☐ 43. Earth-Grazing Meteor
- ☐ 44. Sparkling Sirius
- ☐ 45. Supernova
- ☐ 46. Nova
- ☐ 47. All of Those Other Satellites
- ☐ 48. Stars on Water
- ☐ 49. Crab Nebula
- ☐ 50. Barnard's Star
- ☐ 51. Eta Aquilae, Heartbeat Star
- ☐ 52. Reddest Star
- ☐ 53. Mercury Transit
- ☐ 54. Orion's Belt
- ☐ 55. Ursa Major Moving Cluster
- ☐ 56. Top 10 Lunar Wonders
- ☐ 57. Galactic Duo, M81 And M82

Acknowledgments

Writing a book always involves a neighborhood's worth of gracious souls. I want to thank the great people at Page Street Publishing for offering the opportunity to write a second book and, in particular, my editor Elizabeth Seise, for her bright and encouraging emails and suggestions on how to make the book better.

My tolerant wife saw more of my back than front these past months, as I sat hunched in a corner of the kitchen researching or tapping away on the keyboard. She fed me, humored me, kept the house clean and provided smart and valuable comments when I occasionally turned around to ask.

I also want to thank the many photographers and artists whose beautiful and informative work illuminates these pages including Sharin Ahmad, John Ashley, Yuri Beletsky, Joseph Brimacombe, John Chumack, Bill Dunford, Alan Dyer, David Fogel, Rick Klawitter, Mirko Lahtinen, Thierry Legault, Scott MacNeill, Chris Marriott, Gary Meader, Tom Nelson, Damian Peach, Jeremy Perez, Sebastian Saarloos, James Schaff, Hunter Wilson and Harald Wochner.

Thanks also to Nancy Atkinson and Fraser Cain for their longtime support and Glenn Langhorst for his friendship.

About the Author

Credit: Lisa Hanson

Bob King loves the night sky and hasn't stopped looking up since he was 10 years old. Born in Chicago, he grew up in nearby Morton Grove. When he was 12, he bought a 6-inch (15-cm) reflecting telescope with money earned from his paper route and spent clear nights studying the night sky from his family's backyard.

King graduated from the University of Illinois at Champaign–Urbana with a degree in German. In 1979, he moved to Duluth, Minnesota, to work as a photographer at the *Duluth News Tribune*. Currently photo editor, he also teaches community education astronomy classes and writes a regular blog about what's up in the sky called Astro Bob (astrobob.areavoices.com). King is also the author of *Night Sky with the Naked Eye* (Page Street Publishing, 2016). King is married with two adult daughters and routinely sacrifices sleep for stars.

Index